Dieter Kind
Hermann Kärner

**High-Voltage
Insulation Technology**

Dieter Kind
Hermann Kärner

High-Voltage Insulation Technology

Textbook for Electrical Engineers

Translated from the German by Y. Narayana Rao

With 193 Figures

Friedr. Vieweg & Sohn   Braunschweig / Wiesbaden

CIP-Kurztitelaufnahme der Deutschen Bibliothek

Kind, Dieter:
High-voltage insulation technology: textbook for electr. engineers / Dieter Kind; Hermann Kärner. Transl. from the German by Y. Narayana Rao. – Braunschweig; Wiesbaden: Vieweg, 1985.
   ISBN 3-528-08599-1
   Dt. Ausg. u. d. T.: Kind, Dieter: Hochspannungs-Isoliertechnik für Elektrotechniker

NE: Kärner, Hermann:

Translation by Dr.-Ing. Y. Narayana Rao, Indian Institute of Technology, Madras

1985

All rights reserved
© Friedr. Vieweg & Sohn Verlagsgesellschaft mbH, Braunschweig 1985

No part of this publication may be reproduced, stored in a retrieval system or transmitted in any form or by any means, electronic, mechanical, photocopying, recording or otherwise, without prior permission of the copyright holder.

Produced by W. Langelüddecke, Braunschweig
Printed in Germany

ISBN 3-528-08599-1

Preface

The insulation of components and devices in high-voltage technology is of vital importance and is often the decisive factor in design and construction. Knowledge of the physical processes which limit the electric strength of an arrangement is essential for economical manufacture and reliable performance.

The development of high-voltage insulation technology since the beginning of this century has produced very different solutions for diverse insulations. Characteristic terminology is frequently used pertaining to specific devices, and this often obscures the clear view of the basic scientific principles. Manifold solutions, as a consequence of different properties of the working materials, also impede the understanding of common ground.

This book is principally intended for students of electrical engineering who, towards the end of their course, aim to pursue the difficult path of applying their fundamentals to practice. Very often one has to be satisfied with a solution comprising rough approximations and simplifications, yet one must not lose sight of the basic principles because knowledge of these and their proper application are prerequisites for further development in virgin technological territories.

In full consciousness of this continuity Prof. Dr.-Ing. E.h. Hans Prinz, our mutual academic tutor at the Technische Hochschule München, trained his students accordingly. We are honour-bound to gratefully acknowledge this in these pages.

Beginning with a brief review of the procedure for the determination of electric fields, in the first section on electric strength we consider breakdown phenomena in solids, liquids and gases, in vacuum and on contaminated insulators. The second section comprises a condensed treatment of the properties of high-voltage insulation materials as well as their testing.

Finally, in the third section we discuss constructional peculiarities in high-voltage technology, and subsequently the design and manufacture of capacitors, bushings, lead-outs, transformer windings and instrument transformers with examples. We conclude with tables and diagrams frequently used in the design of high-voltage insulations. Preference has been given to fundamentals rather than detail; a comprehensive bibliography is appended for further assistance where our treatment may be considered rather terse.

The subject material was collated during the many years work as lecturer at the Technische Universität Braunschweig, from the consolidation of personal industrial experience and from research work done at the Braunschweig High-Voltage Institute.

This is an updated version of the German book "Hochspannungs-Isoliertechnik" published 1982, to which many co-workers contributed. We specially thank Dr.-Ing. B. Fell and Mr. J. Nissen for their assistance in revising the German manuscript before translation. We also wish to extend our thanks to a number of colleagues at universities and in industry for the specific perusal of and comment on individual sections.

This English volume was prepared in excellent collaboration with our colleague Dr.-Ing. Y. Narayana Rao, Indian Institute of Technology, Madras, who undertook the translation out of the fund of his own extensive experience in teaching and research. His work was supplemented by Mrs. C. C. J. Schneider M.A. (Cantab.) who carefully revised the English manuscript. Thus the same team was engaged as in the case of the earlier volume "An introduction to High-Voltage Experimental Technique", published 1978 in the same series. The revised manuscript was painstakingly reyped by Miss G. Bosse. Finally we wish to thank the publishers Vieweg-Verlag for their efficient cooperation.

Braunschweig, February 1985

Dieter Kind
Hermann Kärner

Contents

Compilation of the most important symbols used X

1 Electric strength 1
1.1 Electric field and breakdown voltage 1
1.1.1 Determination of electric fields 1
1.1.2 Maximum field strengths in geometrically similar configurations 6
1.1.3 Formulation for the calculation of the breakdown voltage 8
1.1.4 Breakdown probability 10
1.2 Breakdown of gases 14
1.2.1 Charge carriers in gases 14
1.2.2 Self-sustaining discharges 18
1.2.3 Breakdown mechanism in a strongly inhomogeneous field 22
1.2.4 Breakdown under lightning impulse voltages 25
1.2.5 Breakdown under switching impulse voltages 28
1.3 Breakdown of solid insulating materials 30
1.3.1 Charge carriers at low field strengths 30
1.3.2 Intrinsic breakdown 31
1.3.3 Thermal breakdown 37
1.3.4 Partial discharge breakdown 41
1.4 Breakdown of liquid insulating materials 47
1.4.1 Electric strength of technical configurations with insulating liquids 47
1.4.2 Breakdown mechanisms 50
1.5 Breakdown in high vacuum 52
1.6 Pollution flashover 54
1.6.1 Development and effect of contamination layers 54
1.6.2 Mechanism of pollution flashover 57
1.6.3 Pollution tests 60

2 Insulating materials in high-voltage technology 62
2.1 Requirements for insulating materials 62
2.2 Properties and testing of insulating materials 63
2.2.1 Electrical properties 63
2.2.2 Thermal properties 63
2.2.3 Chemical properties 70
2.3 Natural inorganic insulation materials 71
2.3.1 Natural gases 71
2.3.2 Quartz and mica 72

	2.4	Synthetic inorganic insulating materials	72
		2.4.1 Sulphurhexafluoride (SF$_6$)	72
		2.4.2 Glass	74
		2.4.3 Ceramic insulating materials	76
	2.5	Natural organic insulating materials	79
		2.5.1 Mineral oil	79
		2.5.2 Paper	81
		2.5.3 Oil-impregnated paper	82
	2.6	Synthetic organic insulating materials	86
		2.6.1 Molecular configuration and polymerisation reactions	86
		2.6.2 Polyethylene (PE)	87
		2.6.3 Polyvinylchloride (PVC)	89
		2.6.4 Polytetrafluorethylene (PTFE)	90
		2.6.5 Epoxy resin (EP)	91
		2.6.6 Polyurethane resin (PUR)	93
		2.6.7 Silicone elastomer	94
		2.6.8 Chlorinated diphenyls	94
		2.6.9 Silicone oil	96
3	**Design and manufacture of high-voltage equipment**		97
	3.1	Structural details in high-voltage technology	97
		3.1.1 Basic arrangement of the insulation system	97
		3.1.2 Measures to avoid intensification of electric stress	101
		3.1.3 Rigid and leak-proof connections to insulating parts	106
		3.1.4 Measures for air sealing oil-insulated devices	108
		3.1.5 Temperature rise calculation of insulation systems	109
	3.2	High-voltage capacitors	111
		3.2.1 Basic configurations	111
		3.2.2 Design of wound capacitors	114
		3.2.3 Types of design	118
	3.3	Bushings and lead-outs	120
		3.3.1 Basic configurations	120
		3.3.2 Calculation of capacitive gradings	123
		3.3.3 Types of design	128
	3.4	Transformer windings	130
		3.4.1 Design factors for magnetic circuits	130
		3.4.2 Assembly and connection of windings	135
		3.4.3 Insulation of high-voltage windings	137
		3.4.4 Impulse voltage performance and winding construction	138
		3.4.5 Types of transformer winding	140
	3.5	Instrument transformers	142
		3.5.1 Inductive voltage transformers	143
		3.5.2 Capacitive voltage transformers	148
		3.5.3 Current transformers	152

Contents IX

Appendix: Tables and diagrams 159
   A.1 Utilization factors for simple electrode configurations 159
        A.1.1 Spherical configurations 159
        A.1.2 Cylindrical configurations 161
        A.1.3 Point and knife-edge configurations 163
        A.1.4 Circular ring configurations 165
   A.2 Electric strength of gas-insulated configurations 168
        A.2.1 Breakdown voltage in the homogeneous field 168
        A.2.2 Breakdown field strength of plate, cylinder and
               sphere electrodes 169
        A.2.3 Breakdown voltage of rod gaps in air 171
   A.3 Properties of insulating materials 173
   A.4 Properties of magnetic materials 177

Bibliography 180

Index 187

Compilation of the most important symbols used

a	spacing, thickness, length
b	mobility, width, length
d	diameter
e	elementary charge
f	frequency
i	current (instantaneous value)
k	Boltzmann constant, factor, parameter
l	length
l_E	mean iron path length
m	mass, transformation ratio, natural number
n	charge carrier density, natural number
p	geometrical characteristic, pressure
q	charge
q_E	effective iron cross-section
r	radius, spacing
s	gap spacing, thickness, standard deviation
t	time
t_a	formative time lag
t_d	breakdown time
t_s	statistical time lag
u	voltage (instantaneous value)
u_k	relative short-circuit voltage
v	volume, velocity, water content
w	number of turns
A	magnetic vector potential, constant, area
B	magnetic flux density, constant, width
C	capacitance
C'	distributed capacitance, referred capacitance
D	electric displacement, diameter
E	electric field strength, modulus of elasticity (Young's modulus)
Ê	electric field strength (peak value)
E_d	breakdown field strength
E_e	inception (onset) field strength
F	force, volt-time area, relative humidity, error
F'	surface force
F_g	filler content referred to total mass
H	magnetic field strength
I	current (fixed value, effective (rms) value)
Î	current (peak value)

Compilation of the most important symbols used

K	constant, factor
L	self-inductance, length
L'	distributed inductance, referred inductance
P	probability, power
P'	power density
Q	charge
R	radius, real part of impedance (resistance)
R_t	thermal resistance
S	current density, apparent power
T	absolute temperature, periodic time
T_0	reference temperature
T_u	ambient temperature
U	voltage (fixed value, effective (rms) value)
Û	voltage (peak value)
U_d	breakdown voltage
U_e	inception (onset) voltage
U_i	ionization voltage
W	energy
W'	energy density
X	reactive part of impedance (reactance)
Z	apparent impedance (impedance)
α	electron ionization coefficient, thermal transition number, parameter
γ	secondary electron emission coefficient, density
$\tan \delta$	dissipation factor
ϵ	dielectric constant (dielectric permittivity)
ϵ_0	electric constant (permittivity of free space)
ϵ_r	relative dielectric constant (relative permittivity)
η	utilization factor
η_e	attachment coefficient of electrons
ϑ	temperature in °C
κ	conductivity
λ	mean free path, thermal conductivity
μ	permeability
μ_0	magnetic field constant (permeability of free space)
μ_r	relative permeability
ν	natural number, running index
ρ	space charge density, specific resistance
σ	variance of the mean, loss increase, surface charge density
σ_s	layer conductivity
φ	electric potential
ω	angular frequency
Θ	current linkage (magnetomotive force, ampere-turns)
ϕ	magnetic flux

1 Electric Strength

Analogous to mechanical strength, electric strength in insulation systems is a particularly important property. If the electric stress exceeds the electric strength, the insulation either partially or completely loses its insulating capacity. Such a loss can be transient or permanent in nature, correspondingly one differentiates between self-healing and non-self-healing insulations.

Even if all the parameters such as pressure, temperature or humidity are held constant, statement of a single breakdown voltage value is in no way sufficient for a complete description of the electric strength of an insulation. Strictly speaking, as a consequence of the statistical behaviour, a separate value of breakdown voltage results for each voltage stress. Nevertheless, in section 1.1, reference will only be made to voltages U and field strengths E. Only in the following sections is some differentiation made during a detailed consideration of the breakdown mechanisms which occur.

In the practical application of insulations much depends on the sure knowledge, above all, of that voltage which just does not lead to breakdown. This withstand voltage must be capable of being determined from the dimensions and properties of the insulating materials used and proved by means of high-voltage tests on the finished setup. In the case of insulation systems for three-phase networks it is very often sufficient if tests are performed with alternating voltage of short duration and impulse voltage.

1.1 Electric field and breakdown voltage

1.1.1 Determination of electric fields [1])

a) Fundamentals

The intensity of the electric stress at any particular point in the dielectric can be described by the time dependent nature of the appropriate electric field strength $\vec{E}$. Calculation of the field geometry is therefore an important prerequisite for dimensioning an insulation.

Maxwell's equations are the basis for the calculation of electromagnetic fields. From the induction law

$$\operatorname{curl} \vec{E} = -\dot{\vec{B}},$$

one obtains, after introducing the vector potential $\vec{A}$ from the defining equation

$$\vec{B} = \operatorname{curl} \vec{A}:$$

$$\operatorname{curl} \vec{E} = -\frac{\partial}{\partial t} \operatorname{curl} \vec{A} = -\operatorname{curl} \dot{\vec{A}}$$

---
[1]) Comprehensive treatment, see e.g. in [*Lautz* 1969; *Prinz* 1969; *Kuffel, Zaengl* 1984]

or
$$\operatorname{curl}(\vec{E} + \dot{\vec{A}}) = 0.$$

According to the rules of vector analysis, the argument must be capable of being written as the gradient of a potential function, which shall be designated φ. Then it follows for the electric field strength:
$$\vec{E} = -\operatorname{grad}\varphi - \dot{\vec{A}}.$$

Only in special cases, as in short-duration mechanisms in spatially extended high-voltage circuits [*Lührmann* 1973, *Ari* 1974], the contribution of the rapidly varying magnetic field, expressed by $\dot{\vec{A}}$, must be taken into account. Usually, even with high operating voltages, the dimensions are not so large that the finite propagation velocity of the electromagnetic mechanism could be significant. One can then neglect the time variation of the magnetic field:
$$\operatorname{curl}\vec{E} = -\dot{\vec{B}} = 0.$$

The well-known relationship for the electrostatic (irrotational) field is obtained from this as:
$$\vec{E} = -\operatorname{grad}\varphi.$$

It follows that the voltage U, as the potential difference between two points 1 and 2, would be independent of the integration path s:
$$\int_1^2 \vec{E}\,d\vec{s} = -\int_1^2 \operatorname{grad}\varphi\,d\vec{s} = \varphi_1 - \varphi_2 = U.$$

In homogeneous isotropic dielectrics, the electric displacement or flux density $\vec{D}$ is proportional to the electric field strength:
$$\vec{D} = \epsilon\vec{E}, \text{ with } \epsilon = \epsilon_0 \epsilon_r.$$

Fields with space charge density ρ are described by:
$$\operatorname{div}\vec{D} = -\epsilon\operatorname{div}\operatorname{grad}\varphi = -\epsilon\Delta\varphi = \rho.$$

From this follows the Poisson equation:
$$\Delta\varphi = -\frac{\rho}{\epsilon}.$$

In practical insulations space charges do not generally occur at such intensity as to effect a noticeable change in the field generated by the electrode charges. Space charge fields in fact need only be reckoned with when the electric stress exceeds critical values, i.e. when the limiting value of the strength has already been reached. Therefore, for the design of insulation systems, as a rule it is the space charge-free field which is conclusive. Here Laplace's equation is valid:
$$\Delta\varphi = 0.$$

Determination of irrotational and space charge-free electric fields can be done by analytical or numerical calculations or also by measurements on models and in the high-voltage setup. In view of the extensive special literature on this, only a brief survey shall be given here.

1.1 Electric field and breakdown voltage

b) Analytical field calculation [*Lautz* 1969; *Prinz* 1969]

Calculation from Maxwell's integral equation

From the continuity condition it follows that the integral over a closed surface A must be equal to the enclosed charge Q:

$$\oint \vec{D} \, d\vec{A} = Q .$$

Applied to the surface of an electrode, this method can be particularly recommended when the shape of the equipotential surfaces is known, as in spherical and cylindrical configurations.

Charge simulation method

By dividing the volume into sufficiently small volume elements dv any space charge can be broken down into a number of point charges $\rho dv = dq$.
At the field point the potential generated by a point charge at a distance r is:

$$d\varphi = \frac{1}{4\pi\epsilon} \cdot \frac{\rho \, dv}{r} .$$

For a known charge distribution one can superpose the fields generated by the elememtary charges with the help of the Coulomb integral:

$$\varphi = \frac{1}{4\pi\epsilon} \iiint \frac{\rho \, dv}{r} .$$

Similarly, each surface charge can also be divided into a number of point charges:

$$\sigma \, dA = dq .$$

For this, analogously, we have:

$$d\varphi = \frac{1}{4\pi\epsilon} \cdot \frac{\sigma \, dA}{r} .$$

A special form of this method is the method of electrical images, where charges of opposite polarity are introduced, mirrored on the electrode surface.

Direct integration of Laplace's equation

The integration constants which appear for the solution of the differential equation

$$\Delta\varphi = 0$$

must be determined from the boundary conditions. In many cases it is useful to transform the problem to a coordinate system appropriate to the electrode geometry.

Conformal mapping [Prinz 1969]

This method is based upon the fact, well-known from the theory of functions, that all functions of the form

$$\varphi(x,y) = \varphi(x + jy) \text{ with } (j = \sqrt{-1})$$

satisfy the two-dimensional Laplace equation

$$\Delta\varphi = \frac{\partial^2 \varphi}{\partial x^2} + \frac{\partial^2 \varphi}{\partial y^2} = 0 \ .$$

Here $\varphi(x,y)$ must be an analytical function, i.e. continuous and differentiable. Many types of plane fields can be calculated with the help of conformal mapping.

c) **Numerical field calculation** [*Prinz* 1969; *Reichert* 1972]

Difference method of finite elements

The given electrode arrangement is overlaid with a grid mesh whose mesh size represents finite geometrical elements and is chosen to conform to the desired spatial resolution. Laplace's differential equation can then be re-written with the help of approximations in the form of a difference equation. As an example, a plane square grid is represented in Fig. 1.1-1. With the aid of Taylor's expansion it can be shown that the potential at the node 0 for sufficiently small grid spacing can be expressed as the mean value of the potentials of the neighbouring points (square cycle formula):

$$\varphi(0) \approx \frac{1}{4} [\varphi(1) + \varphi(2) + \varphi(3) + \varphi(4)] \ .$$

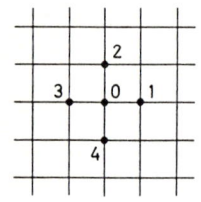

Fig. 1.1-1
Square grid for the difference method

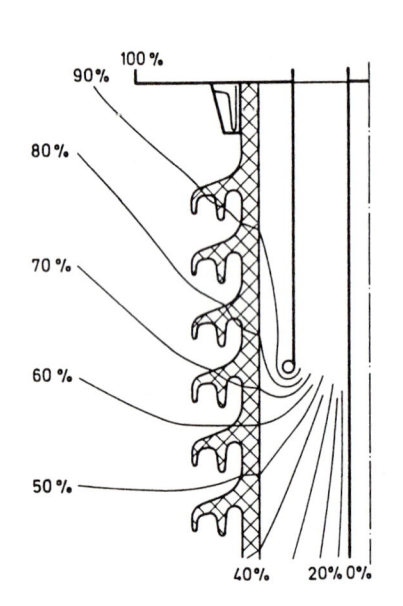

Fig. 1.1-2
Section of the field configuration of a 275 kV current transformer

For the solution, a potential distribution is first assumed and the potentials of the nodes varied in an iterative manner until the boundary conditions on the electrodes for example, and in certain cases, at the boundary surfaces of different dielectrics, are fulfilled to satisfaction.

This method is particularly suitable for spatially restricted field problems with single or multi-material dielectrics [*Shortley* et al. 1947; *Southwell* 1949; *Binns, Lawrenson* 1963; *Galloway* et al. 1967; *Knörrich, Koller* 1970]. Fig. 1.1-2 shows a section of the field configuration of a gas-insulated current transformer for 275 kV [*Ryan* et al. 1971].

1.1 Electric field and breakdown voltage

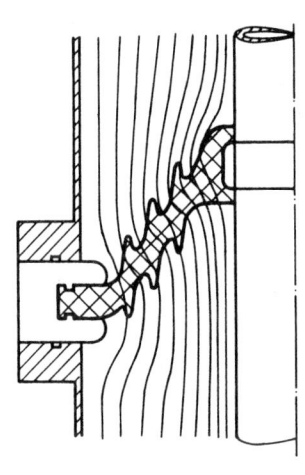

Fig. 1.1-3 Field configuration of the support insulator of a 400 kV tubular busbar

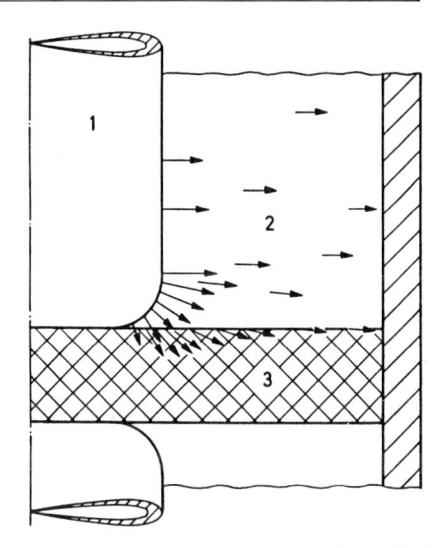

Fig. 1.1-4 Vectors of the electric field of a coaxial symmetrical support arrangement
1 metal, 2 gas, 3 epoxy resin

Fig. 1.1-3 reproduces the potential distribution in the region of a support insulator of a 400 kV tubular busbar, insulated with SF_6 [*Baer, Lehmann* 1974].

Field energy method of finite elements

As in the case of the difference method, the field region to be calculated is overlaid with a grid mesh. With the aid of variational calculus the solution of Laplace's differential equation can be obtained by the minimization of the total field energy of the electrode arrangement. This method is also suitable for field calculation in space charge-free single or multi-material dielectrics [*Zienkiewicz* 1968; *Andersen* 1973; *Böcker, Reichert* 1973; *Weiss* 1974].

Charge simulation method

As already described for the analytical method, the potential at a particular field point is made up of the superposition of the partial potentials generated by the various charges.

In numerical calculation one can now assume the position of equivalent charges on the insides of the electrodes and on the boundary surfaces of different dielectrics and calculate their intensity such that all boundary conditions are satisfied sufficiently accurately.

This method is especially suitable for the exact calculation of the field strength at electrodes in extended setups [*Abou-Seada, Nasser* 1968; *Steinbigler* 1969; *Singer* 1969; *Weiß* 1969; *Singer* et al. 1974; *Singer* 1974]. Fig. 1.1-4 shows, as an example, the field vectors at selected points of the electric field of a coaxial symmetrical support arrangement [*Specht* 1977].

d) Experimental determination of fields [*Philippow* 1966; *Kind* 1972]

On the basis of exact analogue relationships, electrostatic fields can be described by means of electric flow fields. Here the electrolytic tank and conductive papers are among the methods used.

Field measurements at high voltages may also be undertaken in space as well as on the surfaces of electrodes and insulators; here probe methods have been developed which also take into account the effect of space charges [*Wilhelmy* 1973]. In view of the availability of extensive literature, we shall refrain from reproducing further details here.

1.1.2 Maximum field strengths in geometrically similar configurations

For investigation of the electric field strength, the highest field strength E_{max} appearing in the field region is often of interest. To simplify the calculation for practical configurations, *A. Schwaiger* in 1922 introduced the utilization factor:

$$\eta = \frac{E_{mean}}{E_{max}}.$$

E_{mean} is the average field strength of the arrangement over the electrode spacing s, which is usually designated as the gap length. For the homogeneous field, in which a dielectric is uniformly stressed, i.e. "utilized" best, the utilization factor amounts to 100%. In an arbitrary arrangement with two electrodes as in Fig. 1.1-5 after integration along the spacing s shown in a dashed line one obtains:

$$E_{mean} = \frac{1}{s} \int_0^s \vec{E}\, d\vec{x} = \frac{U}{s}.$$

For the applied voltage U the desired maximum field strength is finally obtained as:

$$E_{max} = \frac{U}{s\eta}.$$

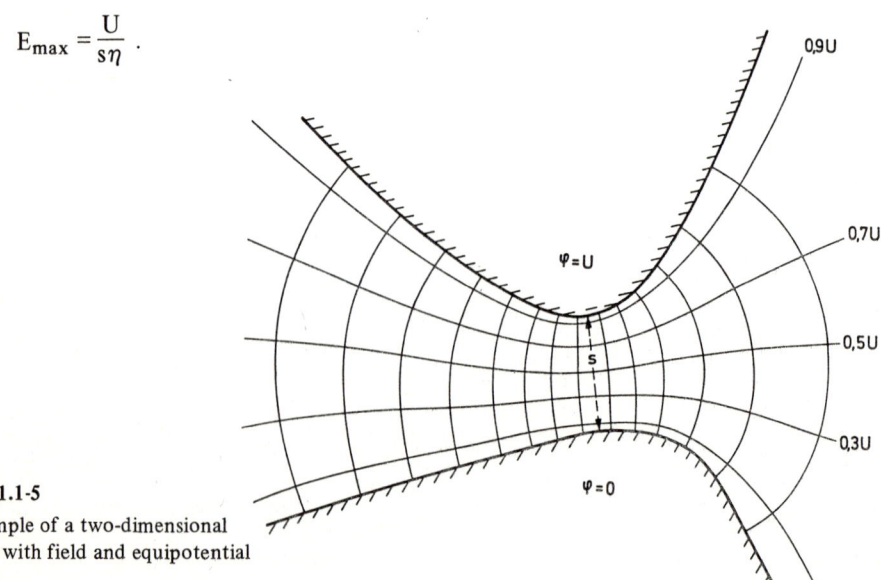

Fig. 1.1-5
Example of a two-dimensional field with field and equipotential lines

1.1 Electric field and breakdown voltage

In geometrically similar arrangements i.e. by variation of the arrangement true to scale, the utilization factor remains unchanged because it depends neither upon the voltage magnitude, nor upon the scale factor. Many electrode configurations can be described unambiguously with the aid of a few geometrical parameters; this is especially true for the practically important spherical and cylindrical arrangements with only one homogeneous dielectric. For these arrangements comprehensive data is available in the literature in the form of diagrams or tables, mostly a result of analytical field calculations [*Schwaiger* 1925; *Morwa* 1966; *Prinz, Singer* 1967; *Prinz* 1969]. Corresponding values of η for a series of two-electrode configurations with a single dielectric are presented in Appendix A 1.

For better understanding, the calculation of η will be reproduced for a particularly simple example. Fig. 1.1-6a shows an arrangement of coaxial cylinders. The highest field strength appears on the inner cylinder and for $s = R - r$ is:

$$E_{max} = \frac{U}{r \ln R/r} = \frac{U}{s} \frac{R-r}{r \ln R/r} = \frac{U}{s\eta}.$$

Fig. 1.1-6
Coaxial cylinder electrodes
a) arrangement
b) utilization factor

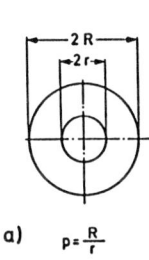

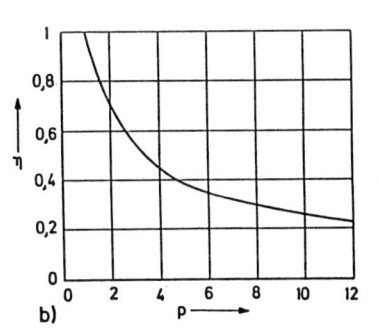

If the parameter p, after Schwaiger, is introduced as a "geometrical characteristic",

$$p = \frac{s+r}{r} = \frac{R}{r}$$

it follows that:

$$\eta = \frac{\ln p}{p-1}.$$

This function is shown in Fig. 1.1-6b.

For a comparative assessment of two-electrode configurations regarding the utilization of the field region, the following rule is valid: for the same spacing s and unchanged electrode at which E_{max} appears, we have the fact that, on account of

$$\epsilon \oiint \vec{E} \, d\vec{A} = Q = CU,$$

the configuration which has the larger capacitance C also has the higher maximum field strength. This is shown in Fig. 1.1-7 for the example of spherical arrangements. It can be seen that η decreases with the degree of mutual envelopment, and the capacitance increases.

The calculation of three-dimensional fields in general requires much more time than the calculation of plane fields. There have been many attempts therefore, to develop approximate solutions for three-dimensional fields by the superposition of two-dimensional

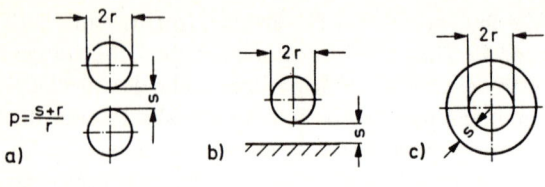

Fig. 1.1-7

Comparison of utilization factor and capacitance for spherical arrangements

	a)	b)	c)	
p	5	5	5	
η	0.372	0.218	0.200	
C/r	0.75	1.28	1.39	pF/cm

fields. Since these methods lack exact theoretical reasoning, one should observe carefully the limits of their applicability [*Boag* 1953; *Prinz, Singer* 1967]. A method often applied to weakly inhomogeneous fields consists in determining, for a certain critical electrode point, the field strengths E_1 and E_2 in those two-dimensional arrangements which result from mutually perpendicular sections through the three-dimensional arrangement. Then for the field strength of the three-dimensional arrangement we have, approximately:

$$E \approx \frac{E_1 E_2}{E_{mean}}.$$

If E_1 and E_2 in both sections are points of highest field strength, for the corresponding utilization factors it follows directly:

$$\eta \approx \eta_1 \eta_2.$$

As proof of the practicability, checking with the values in Appendix A1 establishes that for weakly inhomogeneous fields, the following is indeed valid:

$$\eta_{sphere} \approx (\eta_{cylinder})^2.$$

1.1.3 Formulation for the calculation of the breakdown voltage

The breakdown voltage of an insulation system is that value of the voltage with a certain time dependence for which the dielectric either temporarily or permanently loses its insulating property by way of a discharge process.

One speaks of a complete breakdown when the dielectric is completely bridged by a discharge channel and the arrangement offers only low electrical resistance. The voltage at which a complete breakdown occurs is denoted

Breakdown voltage U_d.

The current flowing subsequently through the insulation is determined solely by the properties of the voltage source; on reaching U_d, depending upon the magnitude and duration of the short-circuit current, a temporary spark discharge or an arc discharge results.

In the case of incomplete breakdown only local overstressing of the dielectric occurs and the insulation continues to determine the current flowing through it. The voltage at which an incomplete breakdown takes place is known as the onset voltage or

Inception voltage U_e.

1.1 Electric field and breakdown voltage

On reaching U_e, spatially limited discharges occur, which, in the event of continuation or increase of the stress, could also lead to a complete breakdown through increasing propagation. The occurrence of these partial discharges and their growth into a complete breakdown depends, for a given stress, upon the geometry of the arrangement, but above all, upon the properties of the dielectric. In homogeneous dielectrics, apart from very short-duration impulse voltages, partial discharges may be expected only in extremely inhomogeneous fields.

For the calculation of U_d and U_e the following concepts are sensible:

a) Assumption of a critical mean field strength $E_{d\,mean}$

Here it is assumed that the breakdown voltage is proportional to the gap spacing:

$$U_d = E_{d\,mean}\,s\,.$$

This simplest assumption of the breakdown voltage proportional to spacing s has its just application for example in large spacings in air (Fig. 1.2-15, curve 1). Of course $E_{d\,mean}$ depends upon many influencing parameters. Values over $100\,kV/cm$ are achieved in solid insulating materials, whilst in air with large spacings the value reaches only about $1\,kV/cm$.

For the inception voltage U_e this concept is not useful, since in a strongly inhomogeneous field U_e varies only little with s (Fig. 1.2-10).

b) Assumption of a critical maximum field strength E_d

In a space charge-free arrangement, the quotient of mean to maximum field strength is independent of applied voltage. Using the nomenclature of Section 1.1.2, it follows:

$$E_{mean} = \eta\, E_{max} = \frac{U}{s}\,.$$

It is assumed that in an insulating material discharges occur when E_{max} reaches a critical value E_d.

It depends upon the arrangement whether for $E_{max} = E_d$ an incomplete breakdown occurs or a complete breakdown results immediately. Using the utilization factor η one obtains Schwaiger's formulation for the calculation of the inception voltage:

$$U_e = E_d\,s\,\eta\,.$$

E_d is designated the breakdown field strength or the breakdown strength and is predominantly a material property, which corresponds to the tensile strength of mechanically stressed materials for example. The spacing s presents itself as a scale factor and η as a scale independent geometrical factor.

In arrangements with weakly inhomogeneous field (say $\eta > 0.3$) or in homogeneous fields a complete breakdown occurs immediately and U_e then becomes equal to U_d. We then have:

$$U_d = E_d\,s\,\eta\,.$$

The field strength E_d not only depends on parameters of the arrangement such as pressure and temperature but also upon the spatial and temporal nature of the electric stress. The limit of applicability of the assumption is approached when E_d also depends to a large extent upon the geometrical parameters, and if the field used for the calculation of η is affected by space charges; this latter always occurs when on reaching U_e a stable incomplete breakdown results (partial discharges, corona discharges).

Despite these limitations, the assumption of a critical field strength in many cases corresponds to the physical data and often supplies sufficiently exact values for U_d and U_e. For liquid and solid insulating materials E_d values of some $100 \, kV/cm$ can be expected and for gases at 1 bar some $10 \, kV/cm$.

The application of both formulations for the calculation presumes a knowledge of $E_{d\,mean}$ or E_d, where these values are only known sufficiently well for specific insulating materials and configurations. For scientific investigation, but also under unusual conditions, such as high temperatures or short stress durations, however, neither of the two formulations can describe the actual breakdown mechanism with sufficient accuracy. In these cases one needs an improved mathematical model for the physical mechanism of breakdown. Sections 1.2 to 1.6 deal with this problem.

1.1.4 Breakdown probability

Physical phenomena can be either exactly reproducible or be subject to statistical scatter. Elastic deformations belong to the first group whereas all destructive mechanisms such as electrical breakdown belong to the second group. During the repeated measurement of a breakdown voltage one should therefore always expect a certain scatter in the values. By applying statistical methods it is possible for relatively few measurements to make statements about the behaviour of an ensemble with determinable certainty (see e.g. [*Kind* 1972]). Moreover for quantitative statements about the electric strength of an insulation the random character of the electrical breakdown too must be taken into account. The random character is also true for the inception voltage and the stress duration. Incidentally, all the considerations of Section 1.1.4 are also valid for the incomplete breakdown and so for the inception voltage U_e.

a) Distribution function

To explain the breakdown probability n identical insulation arrangements shall be considered which one after the other are subjected to a certain voltage stress. This voltage stress will be characterized by the magnitude and the temporal nature of the voltage; the number of breakdowns determined n_d will depend upon the voltage magnitude U. The observed breakdown probability is n_d/n and is a function of U. Fig. 1.1-8 shows the expected form of the distribution function for the electrical breakdown:

$$P(U) = \frac{n_d}{n}.$$

Fig. 1.1-8
Distribution function P(U) of the breakdown voltage

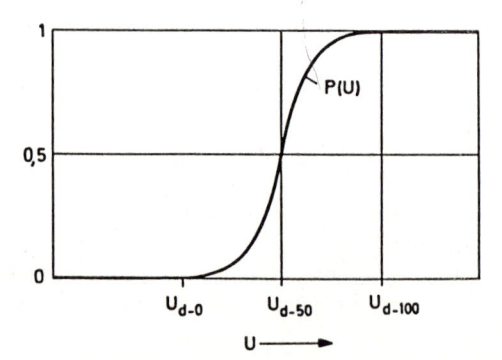

1.1 Electric field and breakdown voltage

Apart from the distribution function P(U), the statistical behaviour of an insulation system can also be described by the distribution density f(U):

$$f(U) = \frac{dP(U)}{dU} \quad \text{or} \quad P(U) = \int f(U) dU .$$

The characteristics plotted in the diagram have the following significance:

U_{d-0} is the withstand voltage, viz. that value of the voltage for which a breakdown is not anticipated (important for all insulations).

U_{d-50} is the median value of the breakdown voltage, viz. that value of the voltage at which, on an average, every second stressing results in a breakdown (important for the measurement of impulse voltages with sphere gaps).

U_{d-100} is the assured breakdown voltage at which breakdowns always occur (important for lightning arresters).

High-voltage insulation systems must be designed so that, at a given stress, the probability of breakdown does not exceed a specified low value.

In technical setups in general, the function P(U) is not known. The values of the breakdown voltage for extreme probabilities can only be approximately determined, since as a rule, the distribution functions approach the limiting values of probability nearly asymptotically. Special arrangements may therefore be necessary to determine U_{d-0} and U_{d-100}.

From n values of U_{di} one can calculate the mean value $\overline{U}_d$ and the standard deviation s according to the following relationships:

$$\overline{U}_d = \frac{1}{n} \sum_{i=1}^{n} U_{di}$$

$$s = \sqrt{\frac{1}{n-1} \sum_{i=1}^{n} (U_{di} - \overline{U}_d)^2} .$$

In practical measurements, the value $\overline{U}_d - 3s$ is often used as the estimator for the withstand voltage U_{d-0} of a setup and $\overline{U}_d + 3s$ for the assured breakdown voltage U_{d-100}. In the event of a normal distribution of the breakdown voltages these limiting values correspond to a breakdown probability of 0.14% or 99.86% respectively.

Experience shows that the actual distributions are well approximated by a normal distribution only in the case of medium probabilities. For very small probabilities, which are significant for insulation systems, exponential distributions named after W. Weibull of the form

$$P(U) = 1 - \exp\left(-\left[\frac{U - U_{d-0}}{U_{d-63} - U_{d-0}}\right]^{\frac{1}{k}}\right)$$

have proved good [*Dokopoulos* 1968; *Artbauer* 1968; *Carrara, Dellera* 1972; *Hylten-Cavallius, Chagas* 1983]. Here k is a constant of the order of 0.1. U_{d-63} is determined from $P(U_{d-63}) = 1 - \frac{1}{e} = 0.632$. The Weibull distribution of the quoted form, compared with the normal distribution, has the advantage that it is described by 3 parameters rather than 2. In this way experimentally determined distributions can be described better. To have the possibility of defining exactly the value of the random variable having the

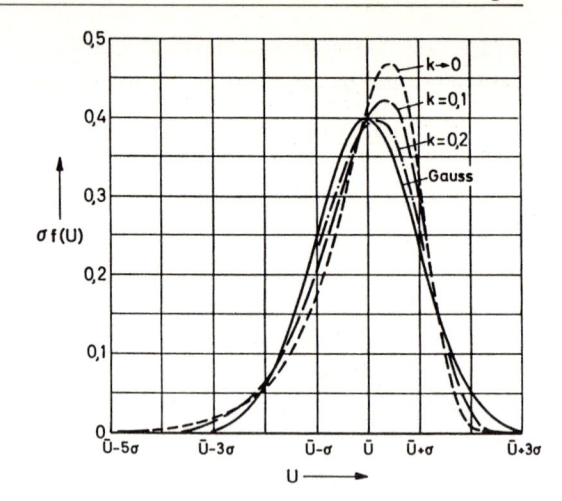

Fig. 1.1-9
Comparison of exponential density functions for various k with a Gauss distribution of the same mean value $\bar{U}_d$ [*Dokopoulos* 1968]

probability P = 0 (withstand voltage U_{d-0}) is particularly important for high-voltage technology (see also b). Fig. 1.1-9 shows a comparison of exponential density functions for various k with a Gauss distribution of the same mean value $\bar{U}_d$ and the standard deviation $\sigma = \lim_{n \to \infty} s$.

b) The law of growth

A further important conclusion to be drawn from the statistical nature of the breakdown is the law of growth of electric strength [*Dokopoulos* 1968; *Widmann* 1964]. It describes the reduction of the breakdown voltage of an insulation system with increasing dimensions, which can also be interpreted as an increasing number of elements stressed in parallel.

To derive the law of growth, m identical insulating arrangements which are statistically independent of one another may be considered (Fig. 1.1-10a). Let the distribution function of each arrangement in itself be $P_1(U)$. The probability that an individual arrangement will not break down is then

$$1 - P_1(U) \, .$$

If m arrangements are stressed simultaneously the probability for no breakdown, according to the multiplication rule of probability theory (equal likelihood theorem), is given by

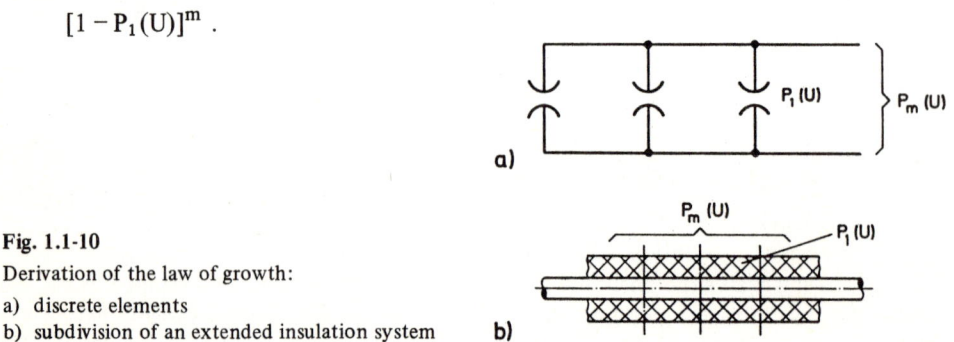

$$[1 - P_1(U)]^m \, .$$

Fig. 1.1-10
Derivation of the law of growth:
a) discrete elements
b) subdivision of an extended insulation system

1.1 Electric field and breakdown voltage

Finally, for the breakdown probability of m parallel arrangements with arbitrary distribution function, it follows that:

$$P_m(U) = 1 - [1 - P_1(U)]^m \, .$$

For $P_1 \ll 1$ this expression reduces to:

$$P_m(U) \approx m P_1(U) \, .$$

Fig. 1.1-11
Distribution function P as a function of the growth factor m for a Weibull distribution with $U_{d-0} = 0$ and $k = 0.2$

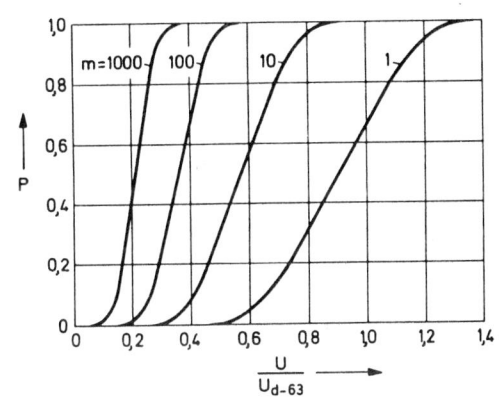

Fig. 1.1-12
Example for insulation coordination
a) circuit
b) form of the distribution functions

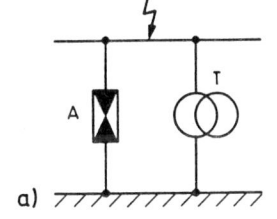

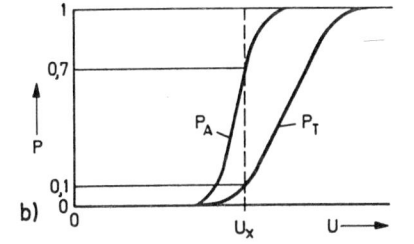

Application of the law of growth to the Gaussian normal distribution leads to fundamental changes in the distribution function, so that the simple evaluation methods of the normal distribution fail. If the concepts leading to the derivation of the law of growth are applied to an extended insulation system such as a cable (Fig. 1.1-10b), a subdivision can only be undertaken arbitrarily. It is only physically meaningful if the overall probability $P_m(U)$ does not vary for any number of elements m. This requirement is satisfied by the Weibull distribution introduced above, since here $P_1(U)$ and $P_m(U)$ are of the same kind. The variation of the Weibull distribution function with the growth factor m is shown in Fig. 1.1-11 for the parameters $k = 0.2$ (corresponding to $\sigma = 0.213 \, U_{d\ 63}$ for $m = 1$) and $U_{d-0} = 0$. The reduction of breakdown voltages for finite probabilities with increasing value of m, also often observed experimentally, is clearly recognizable.

c) Coordination of insulation

The distribution function of the breakdown voltage is of great significance to the economical design of the insulation of high-voltage networks. Since it is not possible to make all the insulations in a setup or network entirely breakdown-proof, the electric strengths of the individual components — expressed by certified voltage tests — must be matched to one another meaningfully. Expensive apparatus such as transformers are designed for a higher test voltage than transmission line insulators.

A simple example for such insulation coordination is shown in Fig. 1.1-12: a transformer T and a lightning arrester A are connected in parallel and their distribution functions are

$P_T(U)$ and $P_A(U)$. On stressing with U_x the lightning arrester would not operate in 30% of the cases; then 10% of these would lead to a breakdown of the transformer insulation. Hence for 100 stressings one should expect $100 \cdot 0.3 \cdot 0.1 = 3$ failures of the transformer.

1.2 Breakdown of gases [1]

The physical mechanisms associated with the development of an electrical breakdown in gases are of principal significance to all types of breakdown. Knowledge of this is therefore a prerequisite for the understanding of breakdown mechanisms in liquid and solid insulating materials too, as well as on boundary surfaces.

Here it is advantageous that gases are usually more homogeneous and more accessible to experiment than liquid and solid insulating materials. Breakdowns in gas-insulated setups can be repeated at short time intervals and with comparatively less scatter; the standard deviation is usually only a few per cent of the mean value. For a wide range of influencing parameters such as pressure, spacing and gas composition, quantitative theories have been developed and verified in numerous measurements.

However, in atmospheric air, the most important gaseous dielectric, appreciable uncertainties exist. The reason lies in the undefined composition, where humidity, dust and charge carriers with very different properties play an important role.

1.2.1 Charge carriers in gases

a) Properties of different charge carriers

In gaseous media, besides the electrically neutral molecules of the gas or vapour component, electrically charged particles can also be present. The most important types of charge carriers are:

Electrons: Charge: $q_e = -e$, with unit charge $e = 1.602 \cdot 10^{-19}$ C
Mass: $m_e = 9.110 \cdot 10^{-28}$ g

Ions: Charge when singly ionised: $q_i = \pm e$
Mass approximately equal to molecular weight M times proton mass:
$m_i \approx M\,2836\,m_e$

Large ions: Formation by attachment of electrons or ions on dust particles, water droplets (fog) and similar macroscopic particles.

Even without the effect of electric fields a number of free charge carriers are usually present in gases. Their density is determined by the equilibrium between new formation and charge carrier loss through recombination and migration. External ionization is generally caused by radiation. Besides corpuscular rays (α and β radiation) it is primarily the short-wave electromagnetic radiation (γ-rays and UV light) that is the cause.

In field-free atmospheric air the ion density generated by cosmic and radioactive radiation corresponds to about 10^3 cm^{-3}. The mean lifetime of an individual ion is about 18 s.

---

[1] Comprehensive treatment in [*Gänger* 1953; *Sirotinski* 1955; *Llewellyn-Jones* 1957; *Loeb* 1960; *Raether* 1964; *Nasser* 1971; *Rein* 1974; *Philippow* 1976; *Mosch, Hauschild* 1978; *Meek, Craggs* 1978; *Kuffel, Zaengl* 1984] and others.

1.2 Breakdown of gases

Electronegative gases have a high electron affinity i.e. a greater inclination for bonding of electrons and so for the formation of negative ions. Oxygen (and hence also air) and above all halogens and halogen compounds as sulphurhexafluoride are gases of this type. The attachment probability is high for electrons of low kinetic energy; electrons of higher kinetic energy can even generate positive ions in electronegative gases. Other gases such as nitrogen, hydrogen and the noble gases have little affinity to form negative ions.

b) Non-self-sustaining discharge

Electrical conduction mechanisms in gases due to movement of charge carriers are described as gas discharge. They can be sustained by means of external influences as irradiation or heating, or even by the discharge itself. In the first case one speaks of a non-self-sustaining discharge and in the second case of a self-sustaining discharge.

The Coulomb force on a point charge q in an electric field $\vec{E}$ is:

$$\vec{F} = q\vec{E} .$$

If a particle with charge q traverses, without collision, a distance x corresponding to a partial voltage ΔU, the increase in its kinetic energy is

$$\Delta W = \int_0^x F(x) dx = q \int_0^x E(x) dx = q \Delta U .$$

In dense gases the movement is disturbed, however, after travelling the free path λ, due to collision with another particle. During collision between heavy particles (ion and gas molecule), an appreciable part of the acquired energy can be transferred to the collision partner; in the case of elastic collision of an electron with a gas molecule the electron would be scattered, retaining nearly all of its total kinetic energy. In the course of numerous elastic collisions or in one non-elastic collision (excitation, ionisation), it can lose a considerable part of its energy. As a consequence of the collisions, for constant field strength, a movement of the charge carrier with constant average velocity results, the drift velocity $\vec{v}$. For small free paths this is nearly proportional to the field strength. The sign corresponds to the polarity of the appropriate charge carrier:

$$\vec{v} = \pm b\vec{E} .$$

In air at normal conditions (1013 mbar, 0 °C) the mobility b introduced here has a value of about:

for electrons $\quad\quad b_e \approx 500 \dfrac{cm/s}{V/cm}$

for positive ions $\quad b_i \approx 1 \dfrac{cm/s}{V/cm}$

for large ions $\quad\quad 10^{-4} ... 10^{-1} \dfrac{cm/s}{V/cm} .$

If one assumes a value for the field strength of 30 kV/cm, then the values of the drift velocity are:

$v_e \approx 150$ mm/μs
$v_i \approx 0.3$ mm/μs .

It follows that for short-duration stresses lasting only a few μs, ions or even charged particles influence the mechanism, not by way of their movement but solely by their presence at a particular location (space charge).

If a homogenous field $\vec{E}$ is applied to a gas which contains electrons of density n_e and singly charged positive ions of density n_i, a carrier current results with current density:

$$\vec{S} = n_i \vec{v}_i e - n_e \vec{v}_e e = (n_i b_i + n_e b_e) e \vec{E} = \kappa \vec{E} .$$

The curve of the current density as a function of field strength shows saturation in the lower region. The saturation current density S_s indicated in Fig. 1.2-1 corresponds to the number of carriers generated by external ionization. Such a weak current dark discharge is non-self-sustaining, since it is entirely dependent on external ionization.

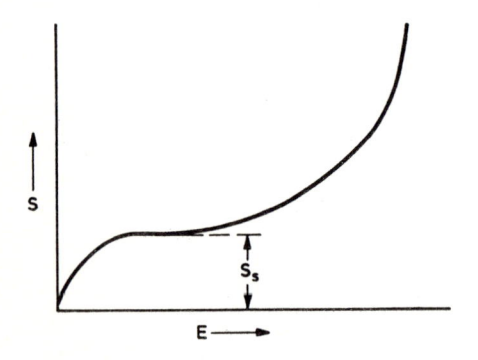

Fig. 1.2-1

Current density of a non-self-sustaining discharge

c) Collision ionization by electrons

For a sufficiently long mean free path and appropriate field strength, electrons can, in collision with a neutral molecule, have such a large kinetic energy ΔW that the molecule is ionized with the release of a further electron. The collision is successful if ΔW attains the ionization energy W_i:

$$\Delta W \geq W_i .$$

The mean value of ΔW can be calculated from the partial voltage $\Delta U = E\lambda$ and the mean free path λ of the electrons; for atmospheric air λ is about $1\,\mu m$. The condition for ionization is often written in the form

$$E\lambda \geq U_i ,$$

where $U_i = W_i/e$ is the ionization voltage of the gas concerned. The following values are valid:

Type of gas	N_2	O_2	Hg	Cs	$SF_6 \rightarrow SF_5^+$
U_i in V	15.8	12.8	10.4	3.9	15.9

1.2 Breakdown of gases

If the ionization condition is fulfilled, an independent multiplication process of the electrons by collision ionization sets in. Depending upon the local field strength, a certain number dn of new electrons is produced over the distance dx:

$$dn = \alpha n(x) dx \; ;$$

$\alpha = \alpha(E)$ is called the ionization coefficient of the electrons.
For a primary electron number n_0 and inhomogeneous field, the integration gives:

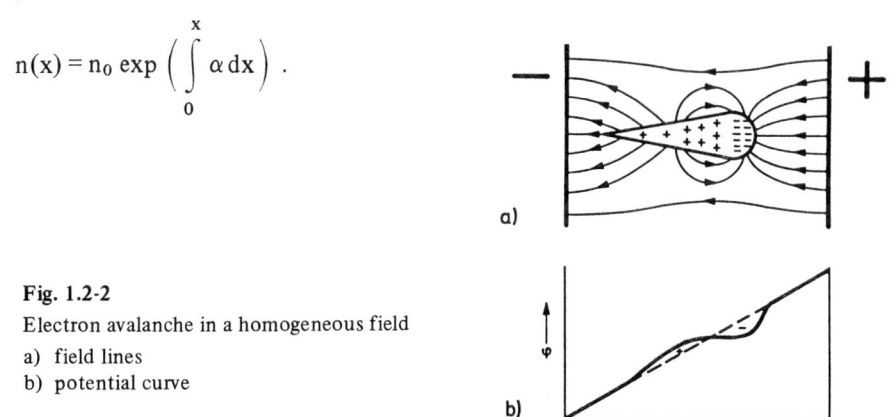

$$n(x) = n_0 \exp\left(\int_0^x \alpha \, dx\right) .$$

Fig. 1.2-2
Electron avalanche in a homogeneous field
a) field lines
b) potential curve

In a homogeneous field, where $\alpha = $ const., this relationship simplifies to:

$$n(x) = n_0 \, e^{\alpha x} .$$

This exponential increase of the electron number is called an electron avalanche. It can be made visible through photography in the cloud chamber. At the head of the avalanche electrons may be very densely packed and, for a high carrier number, can cause great concentration of field lines. Behind the head of the avalanche the positive ions remain (Fig. 1.2-2). These move towards the cathode where under certain conditions they may liberate secondary electrons.

The real free paths λ_ν of the individual electrons constitute a statistical distribution; λ is the mean value by definition. The probability that an electron of mean free path λ will travel a distance greater or equal to λ_ν may be taken to be $\exp(-\lambda_\nu/\lambda)$.

Ionization by a single electron only occurs when $\lambda_\nu \geq U_i/E = \lambda_i$. Over a distance x there are an average x/λ collisions, but the number of ionizations is a factor $\exp(-\lambda_i/\lambda)$ smaller.

The ionization coefficient α can then be calculated as:

$$\alpha(E) = \frac{1}{x}\left[\frac{x}{\lambda} \exp(-\lambda_i/\lambda)\right] = \frac{1}{\lambda} \exp(-U_i/E\lambda) .$$

Further, since the mean free path λ is inversely proportional to the pressure p at constant temperature, we have:

$$\frac{\alpha(E)}{p} = A \exp(-B \, p/E) = f\left(\frac{E}{p}\right) .$$

This relationship characterises the ionization behaviour of a gas.

For air an approximate expression for $\alpha(E)$ is:

$$\alpha(E) \sim (E - E_0)^2 ,$$

where E_0 represents a constant.

d) Attachment of electrons

In electronegative gases (e.g. SF_6) negative ions are formed by the attachment of electrons on neutral molecules. In this way the collision process over the distance dx is deprived of

$$\eta_e \, n(x) \, dx$$

electrons; $\eta_e = \eta_e(E)$ is called the attachment coefficient of the electrons. By introducing the effective ionization coefficient

$$\bar{\alpha} = \alpha - \eta_e$$

into the calculation, this leads to:

$$dn = \bar{\alpha} \, n(x) \, dx .$$

This should be taken into account when dealing with electronegative gases.

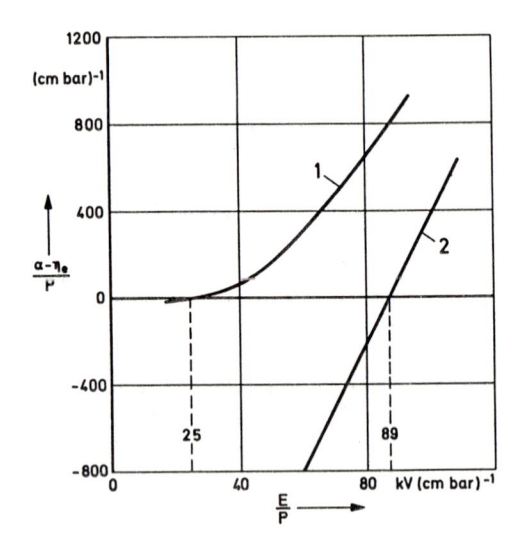

Fig. 1.2-3
Effective ionization coefficient of air (1) and SF_6 (2) at 20 °C

The effective ionization coefficient $\bar{\alpha} = \alpha - \eta_e$ is decisive for the production of charge carriers. In air η_e is very small, the measured curve in Fig. 1.2-3 approximately corresponds to the exponential relationship. For electronegative gases the attachment coefficient η_e has great effect; the practically linear plot was measured for SF_6.

Fig. 1.2-3 shows that $\bar{\alpha}$ of air becomes positive at about $E/p = 25$ kV/cm bar i.e. above this value a positive balance in charge carriers occurs. For SF_6, on the other hand, one would only expect an electron avalanche to be created above 89 kV/cm bar. This is the reason why technical setups in SF_6 have a two to three times greater breakdown strength than in air at the same pressure.

1.2.2 Self-sustaining discharges

a) Townsend mechanism

The voltage on an electrode arrangement in gases can only be increased to the point where a breakdown occurs i.e. a changeover to a self-sustaining discharge. This differs

1.2 Breakdown of gases

from the non-self-sustaining discharge in that the charge carriers required along the path are created by the mechanism itself rather than liberated by external ionization. According to Townsend, ignition of a gas discharge in a homogeneous field can be explained by the fact that new electrons are produced by secondary emission at the cathode. Secondary electrons can be created by incident ions or photons (at pressures over about 10 mbar). If the supply via positive ions is the prevailing process then the number of secondary electrons n_- will be proportional to the number of positive ions n_+ incident on a certain cathode region:

$$n_- = \gamma n_+ .$$

γ is called the secondary emission coefficient. Depending on the experimental conditions, γ can assume values in the range $10^{-8} \ldots 10^{-1}$.

The Townsend model is based on the assumption that an avalanche, produced by n_0 primary initial electrons in the vicinity of the cathode, on traversing the spacing s generates a total of $n_0 (e^{\alpha s} - 1)$ ions. On striking the cathode, by secondary emission these release

$$\gamma n_0 (e^{\alpha s} - 1)$$

secondary initial electrons (Fig. 1.2-4). If this number is larger than the original initial electrons n_0, the current in the configuration increases rapidly without any external assistance, i.e. the gap breaks down. The condition for ignition follows directly from this:

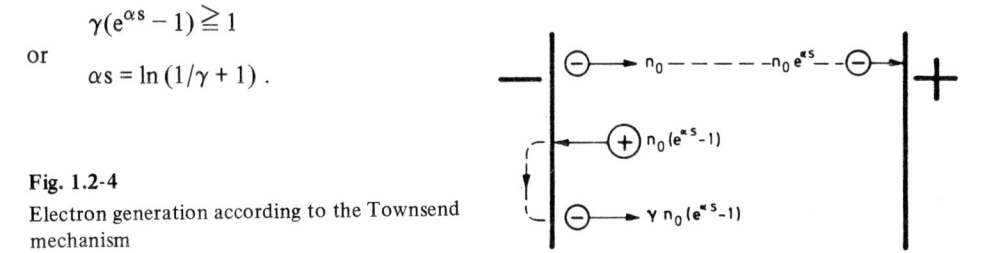

$$\gamma (e^{\alpha s} - 1) \geq 1$$

or

$$\alpha s = \ln (1/\gamma + 1) .$$

Fig. 1.2-4
Electron generation according to the Townsend mechanism

The right-hand side of the equation hardly changes in the usual ranges of γ, and so the ignition condition for a homogeneous field can also be written as [*J. S. Townsend* 1901]:

$$\alpha s \geq k .$$

The values for k lie within the range $k = 2.5 \ldots 18$. The analogous derivation for the inhomogeneous field gives [*W. O. Schumann* 1922]:

$$\int_0^s \alpha \, dx \geq k .$$

In both cases α depends on the field strength E, whose value on attaining the ignition condition is known as the breakdown field strength E_d. The breakdown voltage U_d, and from this E_d of the configuration, can be determined from these equations provided $\alpha = \alpha(E)$ and $E = E(x)$ are known. Here for an inhomogeneous field, the integration must be done along the expected breakdown path; this is the line of force along which the above integral yields the highest value. If the ignition condition is already fulfilled for x < s, then at least the inception voltage U_e is attained [e.g. *Zaengl, Nyffenegger* 1974]. It depends on the configuration of the field whether stable pre-discharges occur.

In electronegative gases the loss of electrons by attachment must be taken into account. Derivation of the ignition condition for the homogeneous field thus leads to the expression [*Mosch, Hauschild* 1978]

$$\gamma(\exp[(\alpha - \eta_e)s] - 1) \geq \frac{\alpha - \eta_e}{\alpha} .$$

Analogous to the above, the ignition condition follows approximately with $\bar{\alpha} = \alpha - \eta_e$ as:

$$\bar{\alpha} \cdot s \geq k .$$

b) Streamer mechanism

On increasing the gap, pressure or potential gradient, the experimental results contradict the conclusions of the Townsend mechanism. Investigations by *Raether, Loeb* and *Meek* have shown that the exponential growth of an avalanche cannot be increased at will since the avalanche becomes unstable at a critical length x_k. Measurements in a homogeneous field show that for atmospheric air

$$\alpha x_k \approx 20 .$$

With the growing number of ionization processes, the original field becomes more and more distorted due to the space charges of the avalanche. On the avalanche front in particular, local increases in field strength occur which are accelerated by the electrons present so that further ionization is facilitated. Moreover, many atoms or molecules are excited so that photons are emitted (Fig. 1.2-5). These light quanta of short wavelength can produce initial electrons in space for subsequent avalanches which, if suitably oriented, may combine with the primary avalanche.

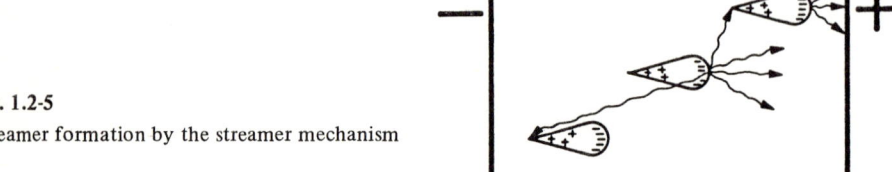

Fig. 1.2-5
Streamer formation by the streamer mechanism

The electrons produced by the radiation migrate from the surroundings towards the still positive area behind the head of the avalanche and a weakly conducting streamer is formed. The forward growth of the discharge is essentially accelerated by the photo-ionization process preceding at the speed of light, and reaches values of $1 ... 10 \, m/\mu s$. As soon as these streamers have established contact between the electrodes, heating-up to a low-resistance plasma streamer generally occurs by the actual breakdown current.

According to the streamer mechanism, a complete breakdown can develop from a single avalanche. The forward growth direction of the streamer here can, because of the range of the photon radiation, even be in the opposite direction to the avalanche. For unsymmetrical arrangements with strong curvature of the positive electrode, cathode-directed streamers of this kind should be expected.

1.2 Breakdown of gases

The critical avalanche length x_k in the inhomogeneous field can be calculated using the relation

$$\int_0^{x_k} \alpha \, dx \approx 20 \, ;$$

streamer discharge only occurs for $s > x_k$. Because of the dependence of α on the field strength, x_k gets much smaller for higher voltage. This explains the very short formative time-lags down to a few ns which have been observed in breakdown testing with strongly overshooting impulse voltages. A quantitative comprehension of the streamer mechanism is not unconditionally possible.

c) Leader mechanism

Very large spacings ($s > 1$ m), as they are unavoidable in outdoor high-voltage installations, usually occur in combination with electrode configurations having a strongly inhomogeneous and frequently unsymmetrical electric field. Particularly when the electrode with the stronger curvature is stressed by a positive switching impulse voltage, a discharge process occurs which is denoted a streamer-leader mechanism. This mechanism leads to bridging of large gaps at comparatively low mean field strength (see Section 1.2.5).

d) Calculation of the breakdown voltage

In Section 1.1.3 various ways of calculating the breakdown voltage U_d of a setup were considered. With a better understanding of the dominant physical mechanism, combined with measurement results, calculation modes can be established which encompass the influence of diverse parameters.

If the relation derived in 1.2.1 c) for the ionization coefficient is substituted into the ignition condition for the homogeneous field valid for the Townsend mechanism, then with $U_d = E_d \, s$ we have:

$$\alpha s = A p \exp\left[-B \frac{ps}{U_d}\right] s = k$$

$$U_d = B \frac{ps}{\ln\left(\frac{A}{k} ps\right)}$$

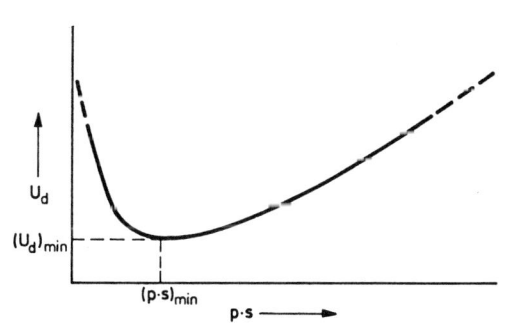

Fig. 1.2-6
Basic shape of a Paschen curve (see also A 2.1)

The most important statement this equation makes, namely that U_d is only a function of the product ps, was discovered experimentally as early as 1889 by *F. Paschen*; this not only holds for the Townsend mechanism but also for the streamer discharge [*Hess* 1976]. Fig. 1.2-6 shows the basic curve shape. In appendix A 2.1 the Paschen curves for various gases are shown. The branch to the left of the minimum is denoted the area of close-range breakdown, the right-hand branch the area of long-range breakdown.

As a consequence of the discharge formation in several avalanche generations, according to the Townsend mechanism, the breakdown times when a step voltage is applied take the values of the ion flight times, namely in the µs range. Since other mechanisms yield even shorter breakdown times, the peak value $\hat{U}_d$ is decisive for alternating voltages and impulse voltages of not too short duration (time to half-value $> 10\,\mu s$).

The Paschen law is satisfied well for static breakdown, i.e. not too rapidly changing stress (in time), empirically up to a certain value of ps. For impulse voltage stress this value decreases with increasing stress gradient [*Koermann* 1955]. For air the upper limit of validity for overvoltages up to 5% above the static value lies at about ps = 10 bar mm. Technical setups in insulating gases thus as a rule lie outside the validity of the Paschen curve. For an almost homogeneous field one may approximate the dependence of the breakdown voltage on pressure p and absolute temperature T as follows:

$$\hat{U}_d \sim \left(\frac{p}{T}\right)^\alpha \quad \text{with } \alpha = 0.7...0.8\,.$$

This dependence on the gas density is valid for the range 1...10 bar [*Mosch, Hauschild* 1978].

The appendix contains in A 2.2 the breakdown field strength $\hat{E}_d$ as a function of the electrode dimensions for plate, cylinder and sphere electrodes in air and SF_6. This is important for a calculation of the breakdown voltage of technical electrode arrangements.

Divergence from the Paschen law occurs not only for too high, but also for too low a pressure p. Below about 10^{-6} bar the laws of vacuum breakdown, of a different nature, take over.

1.2.3 Breakdown mechanisms in a strongly inhomogeneous field

On electrodes with small radius of curvature, as in points, edges or wires, an appreciable increase of the electric field strength takes place, so that the breakdown field strength E_d can be locally attained. The electrons and positive ions generated by collision ionization when the inception voltage U_e is exceeded, move away from the site of their generation as a result of the Coulomb forces acting on them; in electronegative gases the electrons can also form negative ions by attachment. Accumulation of charge carriers of one polarity generates space charge fields which can greatly change the electric field of the configuration.

a) Incomplete breakdown

The mechanisms in a typical arrangement of a positive point against an earthed plate are represented qualitatively in Fig. 1.2-7. The electrons formed in front of the point by collision ionization are drawn away by the anode. A positive space charge remains which reduces the field at the point. For direct voltage this state can remain stationary without a complete breakdown resulting. When the voltage is increased further short duration "brush discharges" dart out from the weakly glowing space charge region. The frequency and range of the brush discharges increase with increasing voltage, until finally there is a complete breakdown at the breakdown voltage U_d.

1.2 Breakdown of gases

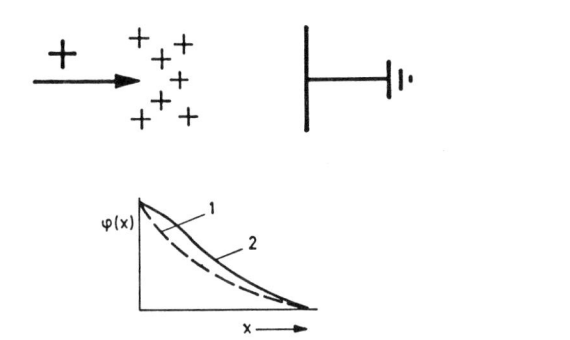

Fig. 1.2-7 Space charges and potential distribution in the case of a positive point
1 without space charges
2 with space charges

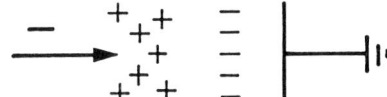

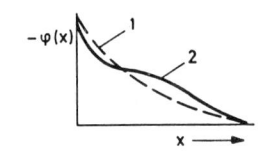

Fig. 1.2-8 Space charges and potential distribution in the case of a negative point
1 without space charges
2 with space charges

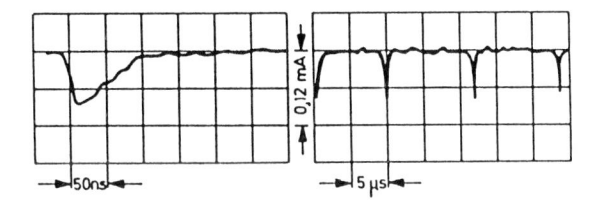

Fig. 1.2-9 Trichel pulses

In the configuration negative point against earthed plate, a rather different behaviour occurs, this is shown in Fig. 1.2-8. Once again a positive space charge results in front of the point when U_e is exceeded, but now the electrons wander in the direction of the plate electrode. If the gas is not able to form negative ions by attachment of electrons, for direct voltage this immediately results in a breakdown since the positive space charge further increases the field strength in front of the point; stationary incomplete discharges are not possible. In most of the technically used gases and especially also in air, however, a space charge consisting of negative ions is formed which can reduce the field in front of the point to such an extent that collision ionization ceases. The discharge sets in once more after the negative space charge has wandered away; the result is a pulse type of mechanism, which leads to regular current pulses of a few 10 ns duration in the external circuit; these were demonstrated for the first time by G. W. Trichel in 1938. Fig. 1.2-9 shows oscillograms of a single Trichel pulse as well as a sequence of such pulses.

With a further increase in the voltage strong current brush discharges occur, even for negative direct voltage and, finally, a complete breakdown results. The rise time of the pulses corresponds to a few ns.

For time varying voltage too, the mechanisms described above occur, although the mechanism of incomplete breakdown is still more difficult to comprehend. For alternating voltage the consequences of periodic polarity changes and, for impulse voltage, the finite duration of the discharge formation must be taken into account.

The discharges occurring during an incomplete breakdown, particularly for alternating voltage, have great technical significance, namely as external partial discharges at points and edges, and corona discharges in overhead transmission lines. In both cases the charge pulses result in high-frequency electromagnetic disturbances which must be taken into account, especially during the design of overhead lines to avoid radio interference in the medium-wave range.

The maintenance of stationary or pulse type discharges, denoted continuous or pulse corona respectively, requires real power. These corona losses in overhead transmission lines depend very much upon atmospheric conditions. Their magnitude lies in the range of about 1...10 kW/km. To achieve a high enough value of the corona onset voltage for overhead lines, the conductor diameter must be chosen to be sufficiently large. For operating voltages above 110 kV, bundled conductors are used instead of single conductors. Three phase transmission lines are designed for a mean r.m.s. field strength of about 15 kV/cm on the surface of the conductor at the rated voltage.

b) Polarity effect during air breakdown

The carriers of positive charge have an appreciably greater mass than the electrons which are the main carriers of the negative charge. For unipolar voltage stresses on unsymmetrical electrode configurations different behaviour should therefore be expected when the electrode with higher field strength changes its polarity.

If one varies the spacing in a sphere-plate configuration in air over a wide range the variation of U_d for direct voltage is as shown qualitatively in Fig. 1.2-10.

For a weakly inhomogeneous field with $s/r < 1$, as in measuring sphere gaps, we have:

$$U_e = U_d \; ; \; U_{d+} \approx U_{d-}.$$

For a strongly inhomogeneous and unsymmetrical field with $s \gg r$ as in the rod-plate configuration, we have:

$$U_e < U_d \; ; \; U_{d+} < U_{d-}.$$

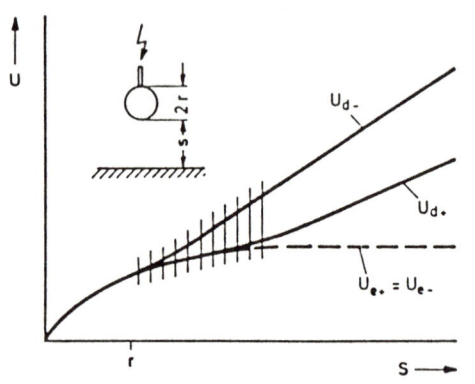

Fig. 1.2-10
Polarity effect in the inhomogeneous field

Large deviations in the breakdown voltage are observed in the boundary region between the weakly and the strongly inhomogeneous field.

The inception voltage U_e, which for large spacings is almost constant and polarity independent, can easily be explained from the nature of the space charge free field; U_e is attained if $E_{max} = E_d$. The occurrence of stable partial discharges at $U_e < U < U_d$ has already been explained on the basis of Figs. 1.2-7 and 1.2-8. What is new is that at large gap spacings the positive breakdown voltage is notably lower than the negative breakdown voltage.

1.2 Breakdown of gases 25

This polarity effect was first observed in 1928 by *E. Marx* under direct and impulse voltages. For alternating voltage the polarity effect always results in the breakdown of unsymmetrical configurations in the positive half-cycle.

A qualitative explanation of the polarity effect can also be obtained from the Figs. 1.2-7 and 1.2-8. The higher breakdown voltage in air for negative polarity of the electrode with the smaller radius of curvature is attributable to the field homogenising effect of the negative ionic space charge.

The breakdown voltage U_d of rod gaps in air is of considerable practical significance to the design of air clearances in equipment and setups for high voltages. From Fig. 1.2-10 it is clear that the radius of curvature r has no appreciable effect on U_d if, due to the effect of space charges, U_d lies considerably above U_e. Actually, all strongly inhomogeneous configurations with gap spacings greater than about 0.5 m behave approximately like rod gaps.

In the appendix, under A 2.3, the dependence of the breakdown voltage $\hat{U}_d$ upon the gap spacing s for alternating and impulse voltages is shown for rod-rod and rod-plate configurations.

1.2.4 Breakdown under lightning impulse voltages

Experience shows that the development of a breakdown requires a finite time. This situation which is important for short-duration stressing, shall be treated here.

a) Statistical time lag

If a voltage which is greater than the inception voltage U_e is applied to a configuration with a homogeneous or only weakly inhomogeneous field, an electron avalanche can only start when an initial electron appears in the critical electrode region. In general, this electron must be generated by natural or artificial external ionization, since field emission requires field strengths of a few MV/cm at the electrodes. The time lag after reaching the breakdown field strength E_d up to the time required for the initial electron to appear, differs from experiment to experiment and is therefore called the statistical time lag.

A simple distribution law for the statistical time lag t_{sv} shall now be derived with the help of a Gedankenexperiment:

On an electrode arrangement consisting of n_0 identical and mutually independent gaps, as in Fig. 1.2-11, let a step voltage $U > U_e$ be applied at $t = 0$. If n is the number of gaps in which no initial electron has yet appeared, the change dn in the time interval dt, with k as proportionality factor, is

$$dn = -kn\, dt\ .$$

The solution is

$$n = n_0 e^{-kt}$$

which is also shown in the figure.

Fig. 1.2-11
Definition of the mean statistical time lag
a) model
b) time variant breakdown distribution

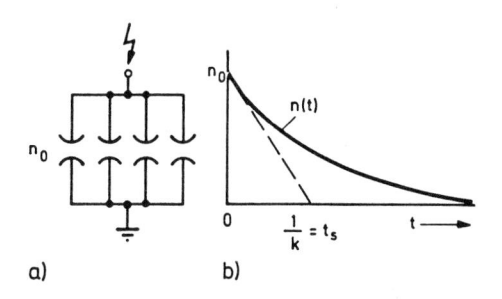

If the experiment is performed on a single gap n_0 times, n then signifies that number of experiments in which the measured time lag $t_{s\nu} > t$. The arithmetic mean value of all the n_0 time lags $t_{s\nu}$ ($\nu = 1 \ldots n_0$) is then given by:

$$t_s = \frac{1}{k}.$$

The mean statistical time lag t_s decreases with increasing size of the electrically highly stressed volume and also with increasing field strength. It is usually only fractions of μs but can in unfavourable cases be some orders of magnitude higher. In the strongly inhomogeneous field an adequate number of charge carriers is made available by the predischarges, so that the statistical time lag has no influence on the development towards complete breakdown.

b) Formative time lag

In breakdown mechanisms what really matters is the movement of charge carriers which are accelerated in the electric field. They move with velocities whose finite magnitude must be taken into account during impulse voltage stressing.

The time interval from the beginning of the first electron avalanche to the formation of the highly conductive breakdown canal, which generally leads to a collapse of the voltage, is designated the formative time lag t_a of the discharge. The processes appropriate to the respective mechanism take place during the time t_a.

The important dependence of the formative time lag t_a upon the magnitude of the applied step voltage U is shown qualitatively in Fig. 1.2-12. If only the static breakdown voltage $U_{d\infty}$ is applied, then very large values result for t_a, while for strongly overshooting voltages, very small values are obtained. Due to uncertainties in the development of the breakdown canal in an inhomogeneous field, even the formative time lag is subject to some scatter. This shall be taken into account in the following by using the designation $t_{a\nu}$.

As a guideline it can be said that the formative time lag in atmospheric air, with 5% overvoltage in a homogeneous and weakly inhomogeneous field, lies well below 1 μs, and above this value in a strongly inhomogeneous field.

c) Impulse voltage-time curves

In an electrically stressed electrode configuration complete breakdown occurs after the combined interval of the statistical time lag $t_{s\nu}$ and the formative time lag $t_{a\nu}$.

The total ignition time lag thus amounts to:

$$t_{v\nu} = t_{s\nu} + t_{a\nu}.$$

Fig. 1.2-12
Voltage dependence of the formative time lag

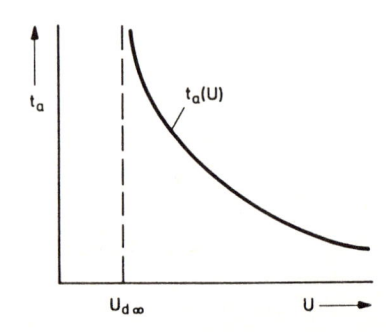

1.2 Breakdown of gases

For impulse voltages with a finite front steepness the ignition time lag is practically counted from that instant at which the static breakdown voltage $U_{d\infty}$ is exceeded. For a complete breakdown to occur, the duration of the stress must be greater than the corresponding ignition time lag. If an electrode arrangement is stressed with a large number of identical impulse voltages u(t) of sufficient magnitude, one obtains pairs of values with breakdown voltage U_d and breakdown time t_d. If the measurement is repeated with impulse voltages of another front steepness, the impulse voltage-time band shown in Fig. 1.2-13 results. It yields the minimum and maximum breakdown times t_d to be expected for a certain configuration at a given impulse voltage. The lower limiting curve 1 of the impulse voltage-time band corresponds to a 0% breakdown probability value, and the upper limiting curve 2 to a 100% value.

For insulation systems, the lower limiting curve is of significance; it is called the formative time characteristic, since here $t_{sv} \approx 0$. Impulse voltage-time curves are a very important basis for dimensioning gas insulation systems stressed with lightning impulse voltages.

For the calculation of impulse voltage-time curves the equal area criterion [*Kind* 1957] has proved to be a useful assumption in many cases. This assumption states that the voltage-time area F between a reference voltage U_b and the formative time characteristic 1 shown in Fig. 1.2-14 remains constant:

$$F = \int_{t_0}^{t_d} [u(t) - U_b] \, dt = \text{const} .$$

For setups with only a weakly inhomogeneous field the reference voltage U_b becomes equal to the inception voltage U_e. If the reference voltage U_b is known, the equal area criterion therefore permits an approximate calculation of the impulse voltage-time curve of an electrode configuration using a few measured values. For different gaps in air and with various types of voltage it has been shown that the equal area criterion permits, with few exceptions, a satisfactory estimation of the volt-time behaviour [*Caldwell, Darveniza*

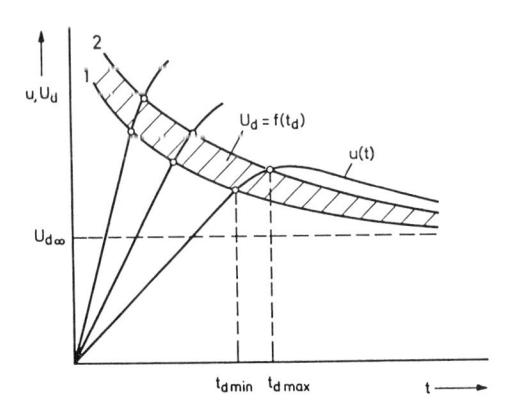

Fig. 1.2-13 Formation and shape of voltage-time curves

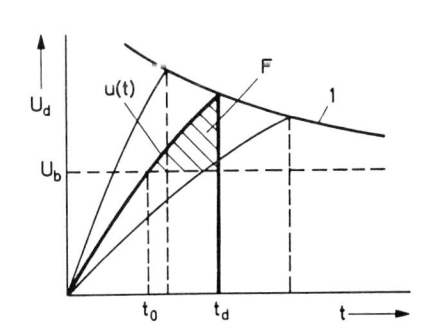

Fig. 1.2-14 Equal area criterion

1973; *Weck, Fischer* 1975; *Waters* in *Meek, Craggs* 1978; *FGH* 1979; *Thione* in *Ragaller* 1980] and its applicability has further been established for weakly inhomogeneous configurations in SF_6 as well [*Knorr* 1977; *Boeck* in *Ragaller* 1980].

1.2.5 Breakdown under switching impulse voltages

Switching impulse voltages, compared with lightning impulse voltages, are characterized by larger pulse duration (e.g. 250/2500 µs according to [IEC Publ. 60-2/1972]). Since they lead to extraordinarily low breakdown voltages for positive polarity of the more strongly curved electrode of inhomogeneous, unsymmetrical electrode configurations with large gap spacings in air, they are the dimensioning parameter for the outdoor insulation systems for 400 kV operating voltage or more. Fig. 1.2-15 points out the differences in strength during stressing of a rod-plate gap with positive lightning impulse voltage or switching impulse voltage.

Whereas for lightning impulse voltage up to the largest gap spacings a constant increase in strength of ca. 5 kV/cm can be relied upon, in the case of switching impulse voltage saturation phenomena occur clearly at gap spacings above 5 m. This is further complicated by the fact that for switching impulse voltage with times to crest $T_{cr} \geq 250$ µs, even lower 50% breakdown voltages occur, whereby the minimum strength for larger gap spacings shifts towards larger times to crest. The air humidity also affects the discharge mechanism and with that the breakdown voltage [*Büsch* 1982]. Thus the curve of minimum strength (lower curve in Fig. 1.2-15) is obtained and must be applied in the design of insulation for highest voltages.

The electric strength of electrode configurations improves, compared with that of the positive rod-plate arrangement, with increasing symmetry and with less inhomogeneity of the field. This phenomenon is accounted for by the introduction of a gap factor k which is defined as

$$k = \frac{U_{d-50 \text{ configuration}}}{U_{d-50 \text{ pos. rod-plate}}}.$$

Fig. 1.2-15
50% breakdown voltage of the rod-plate gap for positive lightning impulse voltage and switching impulse voltage in air (1013 mbar, 20 °C)
1. Lightning impulse voltage 1.2/50
2. Switching impulse voltage 250/2500
3. Curve of minimum strength

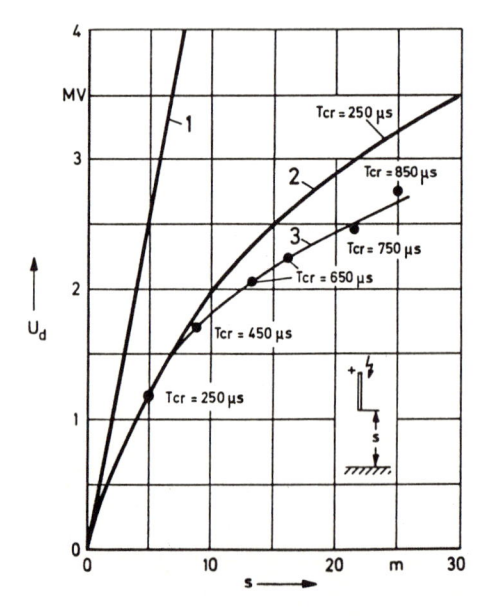

1.2 Breakdown of gases

Since the rod-plate gap shows the lowest strength for positive switching impulse voltage, it has k = 1. Practical electrode configurations have a gap factor of k = 1...2. Since the dependence of the 50% breakdown voltage of a rod-plate gap upon the spacing, shown in Fig. 1.2-15, is known very precisely, the breakdown voltage of any desired configuration can easily be determined without expensive experiments, provided its gap factor is known. An attempt has therefore been made to determine the gap factor from the calculated values of the field of the configuration or those determined on a model (electrolytic tank) [*Schneider, Weck* 1974].

The reason for the low breakdown strength of a rod-plate gap stressed with positive switching impulse voltage should be sought in the type of discharge development, viz. the streamer-leader mechanism. According to Fig. 1.2-16 [*Lemke* 1977] on reaching the ionizing field strength, a streamer discharge develops at the positive point which leaves a positive space charge behind it in the field area. After a dark interval, another more powerful streamer discharge appears under the influence of the increasing voltage and this manifests itself as the first one, by a current pulse. Successive streamer discharges at certain time intervals lead to such high current densities in the vicinity of the point that thermal brush discharges are formed which finally change into the continuously forward-growing leader. From the tip of the leader streamer discharges continue to grow and their current requirement supports thermal ionization and so the existence of the leader. Breakdown is initiated when the streamer reaches the plate electrode.

While the streamer has a voltage requirement of about 4.5 kV/cm, only a voltage of ca. 1 kV/cm drops along the leader. Thus the leader extends the potential of the point into the field area and prepares for its further growth by streamers at the head. In this way the

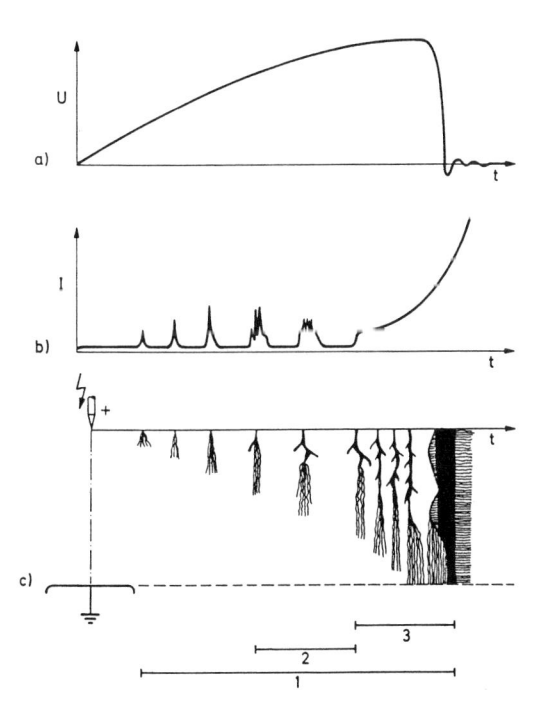

Fig. 1.2-16
Development of the breakdown according to the streamer-leader mechanism
a) voltage curve
b) current curve
c) space-time development of the discharge

Discharge regions:
1 streamer
2 brush
3 leader

gap can be bridged in the manner of intermittent steps. Through chance occurrences in the spatial development of the canal, large scatter in the breakdown voltage can follow. The leader mechanism also explains why the breakdown voltage increases only marginally for an increase in the gap spacing.

Basic research work in this area has been undertaken by an international research group [*Renardières Group* 1974, 1977].

1.3 Breakdown of solid insulating materials[1])

As a rule, solid insulating materials are an inevitable component of every technical insulation system since they necessarily have to act as the mechanical support of the electrodes against one another. Their properties, as may be expected from their number, are very different. Inorganic and organic materials, obtained from natural materials or synthetically manufactured, are those in question; special and growing significance is accorded the synthetic organic materials such as polyethylene (PE) or epoxy resin (EP) mouldings.

A prerequisite for the application of solid insulating materials is sufficient electric strength. For most applications it is important that during a period of time covering decades, high stress is withstood without breakdown.

In the following sections a basic discussion of the electric breakdown shall be considered first. An important parameter for the electric strength of insulating materials is the temperature. In this connection the heat development and temperature distribution have acquired critical significance for the insulation system and the generation of thermal breakdown. The occurrence of partial discharges, which can lead to breakdown as a result of wear, is also a potential danger to the insulation system.

1.3.1 Charge carriers at low field strengths

Free charge carriers can be present in a solid dielectric as positive and negative ions and as electrons. Under the simplifying assumption of only one type of ion, for the current density $\vec{S}$ the relation derived for gases in Section 1.2.1 holds:

$$\vec{S} = (n_i b_i q_i + n_e b_e e)\vec{E} = \kappa \vec{E}$$

with

$$\kappa = \kappa_i + \kappa_e .$$

Here for ions (index i) and for electrons (index e) n denotes density, b mobility, q charge, and κ the specific conductivity.

As in insulating liquids, ionic conduction is due to movement of ions formed by the dissociation of electrolytic impurities or ageing products. In materials with an ionic lattice, at higher temperatures thermally activated lattice ions should also be expected. The corresponding part of the specific conductivity κ_i increases strongly with the temperature due to the increasing degree of dissociation and the growing mobility. We have:

$$\kappa_i = A e^{-\frac{W}{kT}} .$$

---

[1]) Comprehensive treatment in [*Whitehead* 1951; *Franz* 1956; *v. Hippel* 1954; *Lesch* 1959, *Anderson* 1964; *Alston* 1968; *Bartnikas, McMahon* 1979] and others.

1.3 Breakdown of solid insulating materials

In this equation W is the thermal activation energy, k is the Boltzmann constant, T is the absolute temperature, and A is a proportionality factor.

The d.c. conductivity observed in crystals or partly crystalline materials like polyethylene is predominantly caused by electron conduction. The permitted energy levels of the electrons in an insulator are represented in Fig. 1.3-1, based on the band model.

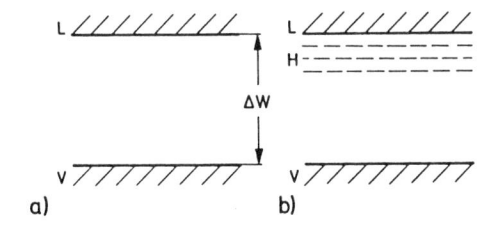

Fig. 1.3-1 Band model
a) ideal insulator
b) insulator with traps

For an ideal insulator as in a), a forbidden zone of width ΔW of a few eV exists between the valency band V and the conduction band L; there are no free electrons in the conduction band L, the conductivity is zero. By perturbation of the lattice structure or due to the presence of foreign atoms, the possibility arises, as in b), for the electrons to populate the forbidden zone at traps H located just beneath the conduction band [*Whitehead* 1951 and *Anderson* 1964]. From there single electrons, without having to cross the entire width of the forbidden zone, can move into the conduction band. Through this internal field emission (tunnel effect), recognised as such in 1934 by *C. Zener*, an appreciable increase in the number of electrons in the conduction band can occur even at room temperature, from which follows the increased electronic conductivity.

In very high-quality electron conducting insulating materials however, estimation shows that an appreciably larger numer of electrons often participate in charge transport than would be expected by thermal activation. The conclusion is that electrons are also injected by external field emission from the electrodes into the solid dielectric [*Mierdel* 1967]. Yet such electrons result in space charge which eventually inhibits the current density.

1.3.2 Intrinsic breakdown

The mechanism which leads to a sudden loss of insulating capability even after a short period of stressing, without appreciable pre-heating and without partial discharges, is called intrinsic breakdown. Here, as a rule, except in the case of strongly inhomogeneous field configurations, a complete breakdown occurs within a few nanoseconds in solid insulating materials. Since this is a short-term mechanism, the relatively small influence of the time dependence of the voltage is neglected in this section; unless otherwise stated, the field strength meant is the highest attained instantaneous value.

Earlier theories assumed an ionic mechanism or mechanical destruction as a consequence of field forces on the lattice ions. However, after the investigations of *A. v. Hippel* in 1931, the work of *H. Fröhlich*, as well as on the basis of recent investigations, one may regard an electron mechanism as being certain [*Franz* 1965]. Intrinsic breakdown must be anticipated in homogeneous, predominantly crystalline materials and especially during short-period stressing.

a) Breakdown of thin plates

At high electric field strength E in a solid insulating material a large increase of the conductivity κ is observed, which must lead to the conclusion that there is an increase of free electrons in the conduction band. Collision ionization as well as increased internal and external field emission can be considered the cause of this. If the field strength continues to be increased, on approaching the breakdown field strength E_d the electron current density reaches such high local values that it leads to heating of the insulating material due to the Joule effect [*Mierdel* 1967]. The power which destroys the dielectric lies in the order of 10^{-5} W/mm^3 [*Thoma* 1976].

If free electrons are accelerated by the electric field as in gases, collision ionization can result. By collision with lattice ions the electrons on giving up their kinetic energy can generate additional free electrons. Under favourable conditions an electron avalanche can thus be created. The formation and growth of the electron avalanche is inhibited by heating as a result of increased deceleration of the electrons by lattice oscillations. In crystalline materials in the low temperature region, it has actually been observed that the breakdown voltage varies only little with increasing temperature, or indeed, even increases (low temperature breakdown, see Fig. 1.3-2, curve 1) [*Franz* 1956].

By the internal field emission mechanism electrons from the valency band or from the traps can take up energy from the field without collision and so move into the conduction band. The process of internal field emission is schematically represented in Fig. 1.3-3 on the basis of the energy band model [*Franz* 1956]. If an insulator 1 of thickness s is arranged between two electrodes 2 and 3, the Fermi levels F of the two electrodes shift

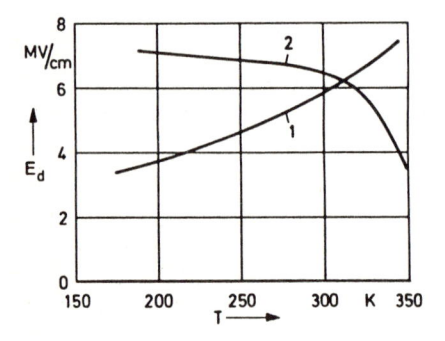

Fig. 1.3-2 Breakdown of quartz under direct voltage
1 quartz crystal, 2 quartz glass

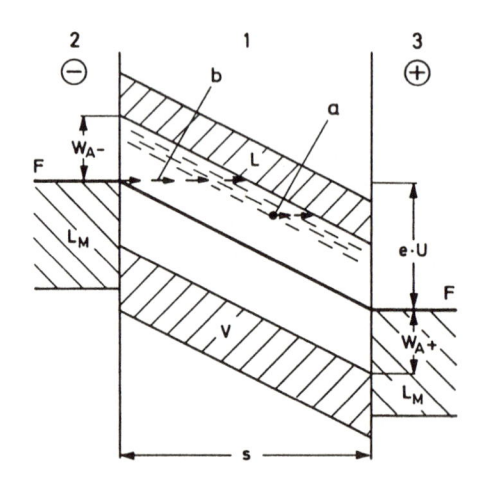

Fig. 1.3-3 Internal and external field emission in the band model
1   insulator of thickness s with conduction band L and valancy band V
2   cathode, W_{A-} electron work function
3   anode, W_{A+} hole work function
L_M conduction band of the metal
⟨F Fermi level⟩

1.3 Breakdown of solid insulating materials

against each other by an amount eU and the energy levels in the insulator are correspondingly inclined. The tangent of the angle of inclination is:

$$\frac{eU}{s} = eE .$$

If the conditions for tunnelling are fulfilled, electrons from the valency band V, yet even easier from the traps (path a), can move into the conduction band L of the insulator.

Internal field emission should be particularly expected in amorphous insulating materials and those containing impurities. When a certain limiting temperature is exceeded a sufficiently large number of traps can be emptied. For example, the decreasing breakdown strength of amorphous quartz with increasing temperature supports the assumption of electron generation by internal field emission (high temperature breakdown, see Fig. 1.3-2, curve 2). Here the following relationship between the breakdown field strength E_d and the absolute temperature T is valid [*Whitehead* 1951; *Anderson* 1964]:

$$E_d \sim e^{\frac{\Delta W}{2kT}} .$$

In polyethylene values between 0.03 eV and 0.05 eV have been determined for the activation energy ΔW [*Luy, Oswald* 1971].

During external field emission electrons are introduced into the dielectric from the cathode, preferentially in the region of local field enhancement. This mechanism is indicated in Fig. 1.3-3 by the transition b. This should be of particular significance in configurations with strongly inhomogeneous field, whereas in the homogeneous field for the region of breakdown field strength, one would expect electron multiplication by internal field emission to predominate [*Mierdel* 1967].

b) Breakdown in technical solid insulation systems

Technical insulation systems differ from those used for fundamental investigations of the physical mechanism according to a). Here the thin layers are replaced by larger wall thicknesses; multi-layer dielectrics and more strongly inhomogeneous electric fields are used. Yet in many cases the developed models can be at least qualitatively applied even to important high voltage insulating materials, especially to plastics [*Luy, Oswald* 1971; *Zoledziowski, Soar* 1972; *Böttger* 1973]. As an example, Fig. 1.3-4 shows measurements of polyethylene foils which demonstrate the expected different behaviour of electric strength at low (predominantly collision ionization) and higher (predominantly internal field emission) temperatures [*Ieda* 1972].

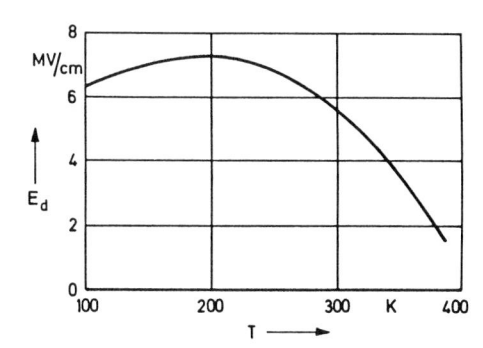

Fig. 1.3-4
Breakdown of PE foil under direct voltage

The boundary temperature between the two regions is shifted towards lower values by the impurities, so that in technical applications one may often expect internal field emission. This is in agreement with the observation that the breakdown voltage usually decreases with increasing temperature [*Artbauer* 1965].

The fact that the breakdown field strength of thin layers below about 100 μm reaches very high values [*Whitehead* 1951], agrees well with the theory. As in gases, the reason for this is that in small electrode spacings the mechanisms for charge carrier multiplication are impeded. For larger insulation material thickness of a few millimetres and correspondingly higher voltages, evidence of the breakdown field strength being independent of the thickness is forthcoming only under favourable conditions. Between 1 MV/cm and 3 MV/cm are the values measured for unfilled EP-mouldings [*Schiweck* 1969; *Schirr* 1974]. Very often, effects originating in the boundary layer between the dielectric and the electrodes dominate, and so in technical configurations there is still a dependence on thickness [*Schühlein* 1968].

As expected from the model concept, the degree of inhomogeneity of an insulating material has great affect on intrinsic breakdown. In general, the material is required to be as homogeneous as possible, since any disturbance of the regular structure would produce traps in consequence, from which electrons can be liberated relatively easily. Similarly, structural boundary surfaces act as weak spots, as for instance in EP-mouldings with quartz powder fillers. Fig. 1.3-5 shows the decrease of breakdown field strengths measured with impulse and alternating voltages for increasing filler content F_g [*Schirr* 1974].

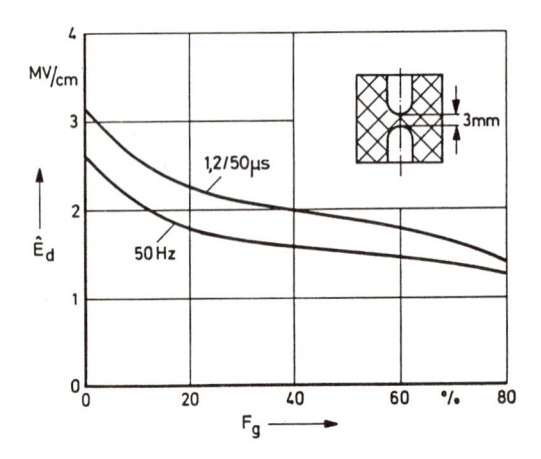

Fig. 1.3-5
Breakdown of EP-moulding test samples under lightning impulse and alternating voltage

Another example for the strength decreasing effect of the structural boundaries is the observed preferential growth of breakdown canals along the spherulite boundaries in the case of partly crystalline polymers [*Menges, Berg* 1972; *Wagner* 1973] (Fig. 1.3-6). Such weak spots which are basically unavoidable in plastic insulations, together with the bonding between the electrodes and the insulating material, determine the breakdown performance of a configuration. The unfavourable effect of external and internal mechanical stresses [*Schirr* 1974; *Jähne* 1975] as well as the thickness and volume effects observed in insulation systems [*Dokopoulos* 1968; *Artbauer* 1968] may be attributed to

1.3 Breakdown of solid insulating materials

Fig. 1.3-6
Micrograph of a discharge canal lying between two polypropylene spherulites
[*Patsch, Wagner, Heumann* 1976]

the existence and the statistical distribution of weak spots. For practical application however, it follows from this that the permissible electric stresses lie well below those values which were determined in laboratories for idealized configurations [*Kind* 1971].

Unfortunately, application of the work on intrinsic breakdown based on the theoretical and experimental considerations of physics, is not yet possible today in the design of practical high-voltage insulation systems. Nevertheless, the models proposed have contributed considerably to technical development.

c) Breakdown mechanisms on electrodes with small radius of curvature

In solid insulation systems one would always aim to avoid electrodes with small radius of curvature. For economical reasons the objective is to utilize the dielectric as uniformly as possible.

But sharp points or sharp edges cannot be excluded with certainty in practical insulation systems. On the other hand, needle electrodes are very often chosen in test objects for breakdown investigations of solid insulating materials and conclusions are drawn from these results about the behaviour of the insulating material under high electric stress [*Bahder* et al. 1974; *Löffelmacher* 1976].

In alternating voltage investigations a constant voltage is usually applied and the onset time t_e is measured, which is the time up to the appearance of the first partial breakdown canal. If stressing is continued, the canals widen as a rule in the form of branches (treeing), until finally after the breakdown time t_d a complete breakdown results. The processes in the time interval t_e to t_d (deterioration period) come under the domain of partial discharge breakdown and shall be discussed in Section 1.3.4. The processes up to t_e (induction period) on the other hand, insofar as short-term mechanisms are concerned, are associated with intrinsic breakdown.

The appearance of the first partial discharge canals on needle electrodes in EP-mouldings has been closely investigated using impulse voltage [*Dittmer* 1963; *Schiweck* 1969]. Fig. 1.3-7 shows the sketch of a canal which can arise due to stressing with a chopped lightning impulse voltage.

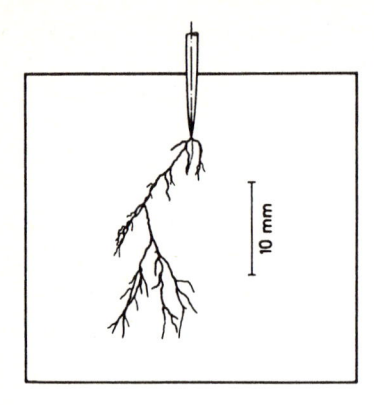

Fig. 1.3-7
Partial discharge canal in unfilled epoxy resin moulding

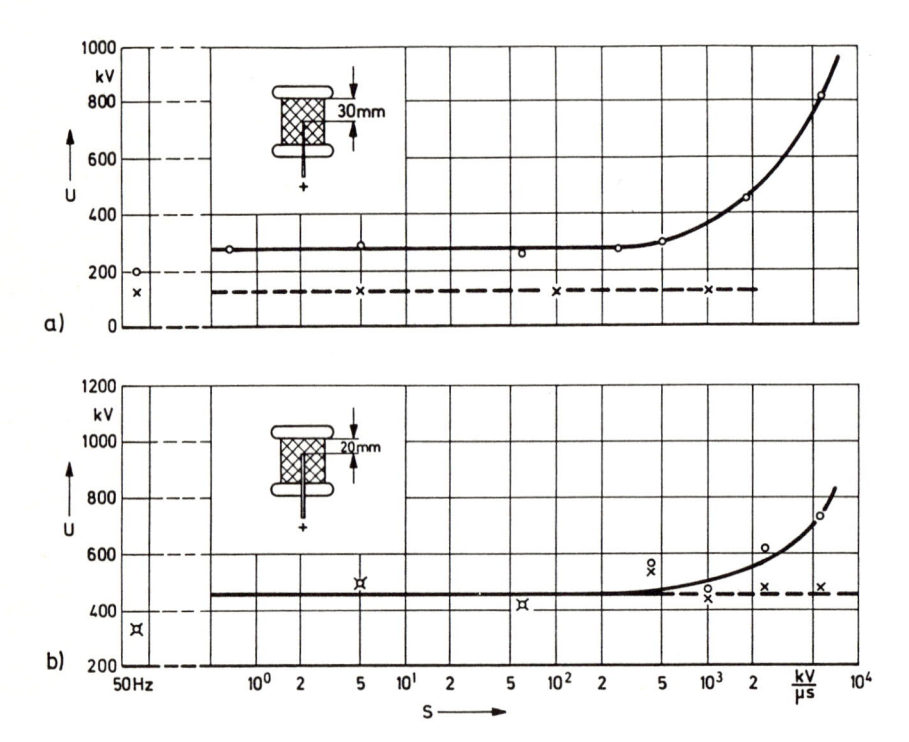

Fig. 1.3-8 Breakdown of EP-moulding test samples (U_d ———; U_e - - - -)
a) needle-plate   b) rod-plate

It has been determined during these investigations that the inception voltage U_e remains more or less constant in the range of a few μm for variation of the needle radius, which suggests field weakening by space charges of the needle polarity. At any event, highest field strengths up to a few MV/mm should be expected, at which field emission becomes possible. With the needle negative, electrons are injected into the insulating material by external field emission and held there in traps by the formation of negative space charge.

1.3 Breakdown of solid insulating materials

With the needle positive, on the other hand, electrons can be liberated from the valency band or from the traps by internal field emission, leaving behind a positive space charge. This mechanism can also be interpreted as injection of positive holes from the anode while overcoming the work function W_{A+}. The field transformation by space charges takes place within a few ns, which offers an explanation of the only very small dependence of U_e on the time curve of the electric stress.

Fig. 1.3-8 shows as an example U_e and U_d of test samples made of unfilled EP-mouldings with needle and rod electrode during stressing with 50 Hz alternating voltage and with positive wedge-shaped impulse voltages as a function of their steepness S for values up to $10\,MV/\mu s$. For negative polarity of the needle, the same onset voltage is measured.

A partial discharge canal is created as a consequence of the avalanche-like increase of charge carriers, provided the increase with time of the field strength, because of the voltage stress, dominates over the field reduction due to space charges. Actually, the occurrence of partial discharge canals at the tip of the needle electrodes has been observed only in the rising part of the impulse voltages, i.e. while the voltage is increasing with time.

The behaviour of configurations with electrodes of small radius of curvature during short-duration stressing, described for the example of unfilled EP-mouldings, agrees with the behaviour of other insulating materials, especially polyethylene. This can be explained with the aid of the model for intrinsic breakdown outlined in a). During long-term stressing with alternating voltage, undertaken in test methods for assessing the suitability of plastic materials (needle test), one must not exclude other mechanisms for the formation of the first partial discharge canal; above all, thermal and electro-mechanical effects should be regarded [*Bahder* et al. 1974].

1.3.3 Thermal breakdown

In insulating materials dielectric losses P_{diel} occur which comprise conduction, polarization and ionization losses. These losses increase the temperature of the dielectric and are themselves temperature dependent. In regions in which the dielectric losses increase steeply with temperature, there is danger of overheating in solid insulating materials, and this eventually leads to breakdown. This fundamental mechanism is named "thermal breakdown" and was described quantitatively for the first time in 1922 by *K. W. Wagner*.

a) Temperature dependence of dielectric losses

For the specific dielectric losses in an alternating field the statement

$$P'_{diel} = E^2 \,\omega \epsilon_0 \, \epsilon_r \tan\delta$$

is valid. The loss factor $\epsilon_r \tan\delta$ is a dimensionless measure of the magnitude of the dielectric losses of an insulating material and its value lies in the region of 10^{-3} to 10^{-1}.

For the d.c. field one obtains:

$$P'_{diel} = E^2 \kappa \ .$$

In both cases for the dependence on temperature T, one may choose the same assumption:

$$P'_{diel} = E^2 \, p(T) \ .$$

Here we have:

for alternating voltage $p(T) = \omega \epsilon_0 \epsilon_r \tan\delta$
for direct voltage $\quad p(T) = \kappa$.

For temperature dependence, the statement

$$p(T) = p_0 \, e^{\sigma(T-T_0)} .$$

has proved useful, where T_0 and p_0 are reference quantities, σ is the loss increase. This relation shall be the basis for the considerations under b).

b) Models to describe thermal breakdown

At first the simple configuration as in Fig. 1.3-9a shall be considered, in the dielectric of which the temperature T and the specific dielectric losses P'_{diel} shall be regarded locally constant. Let the thermal conduction via the electrodes 1 and 2 with the cooling power P_{ab} be proportional to the difference over the ambient temperature T_u:

$$P_{ab} \sim (T - T_u) .$$

A stable operating point must satisfy the following conditions (Fig. 1.3-9b):

$P_{ab} = P_{diel}$ $\quad$ as a prerequisite for static conditions

$\dfrac{dP_{ab}}{dT} > \dfrac{dP_{diel}}{dT}$ $\quad$ as a prerequisite for stability.

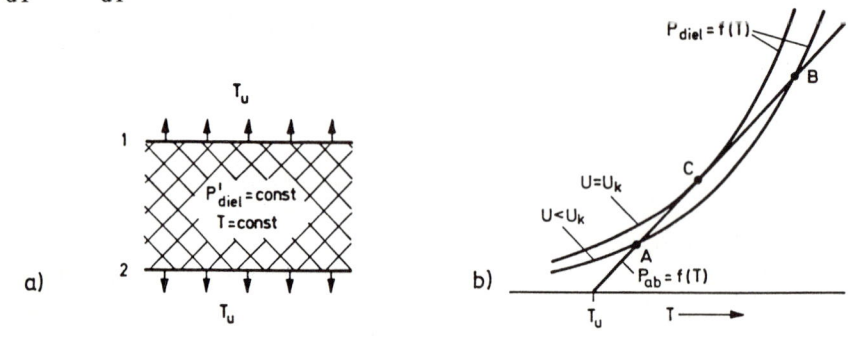

Fig. 1.3-9 Model for the qualitative explanation of thermal breakdown
a) configuration $\quad$ b) characteristics

Thermal breakdown sets in if there is no stable operating point. It is clear that only the point of intersection A can be a stable operating point, whereas B is unstable. By increasing T_u or by raising the voltage U, the points A and B can finally merge at C. The corresponding voltage is designated the critical voltage U_k; it is the voltage of thermal breakdown.

For the qualitative consideration as in Fig. 1.3-9, a locally constant temperature in the dielectric was assumed. For quantitative investigation of the breakdown performance, however, the temperature distribution in the insulating material must be taken into account. This improved model of a plate in a homogeneous electric field is shown in Fig. 1.3-10. This model is based on the assumption of constant temperature T_u of the

1.3 Breakdown of solid insulating materials

electrodes 1 and 2 and therefore thermal conduction in the x-direction only; in addition the thermal conductivity λ of the insulating material is assumed to be constant. With maximum temperature T_m at $x = 0$ the boundary conditions are:

$$x = 0; \quad T = T_m \text{ and } \frac{dT}{dx} = 0,$$

$$x = \frac{s}{2}; \quad T = T_u.$$

Fig. 1.3-10
Model for the calculation of the critical voltage of a global thermal breakdown

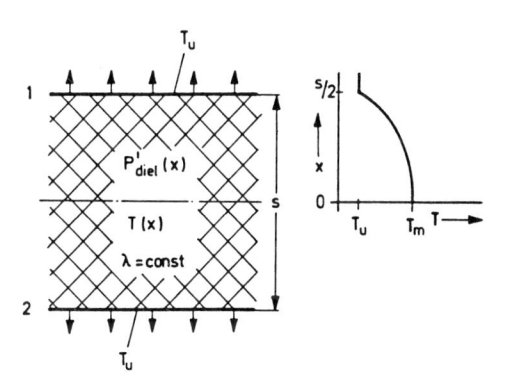

In the static case, for each volume element the power transported away by thermal conduction

$$P'_{ab} = -\text{div } \lambda \text{ grad } T$$

must be equal to the power input P'_{diel}.

For the present one-dimensional problem we have, using the quantities introduced under a), the following differential equation:

$$\lambda \frac{d^2T}{dx^2} + E^2 p_0 e^{\sigma(T-T_0)} = 0.$$

To solve, we rewrite in $\frac{dT}{dx}$ and the following expression is obtained:

$$\lambda \frac{d}{dx}\left(\frac{1}{2}\left(\frac{dT}{dx}\right)^2\right) + E^2 p_0 \frac{d}{dx}\left(\frac{1}{\sigma} e^{\sigma(T-T_0)}\right) = 0.$$

The constant of the integration from $x = 0$ to x is obtained from the boundary conditions and we have:

$$\frac{\lambda}{2}\left(\frac{dT}{dx}\right)^2 + \frac{E^2 p_0}{\sigma}\left(e^{\sigma(T-T_0)} - e^{\sigma(T_m-T_0)}\right) = 0.$$

Separation of variables leads to the integral:

$$\int_{T_m}^{T_u} \frac{dT}{\sqrt{1 - e^{\sigma(T-T_m)}}} = E\sqrt{\frac{2 p_0}{\lambda \sigma}} e^{\frac{1}{2}\sigma(T_m-T_0)} \int_0^{s/2} dx.$$

The integral on the left hand side is solved by substitution. If we write $U = sE$, we finally obtain the relation between voltage and maximum temperature:

$$U = 2\sqrt{\frac{2\lambda}{p_0 \sigma}} \frac{\cosh^{-1} e^{\frac{1}{2}\sigma(T_m-T_u)}}{e^{\frac{1}{2}\sigma(T_m-T_0)}}.$$

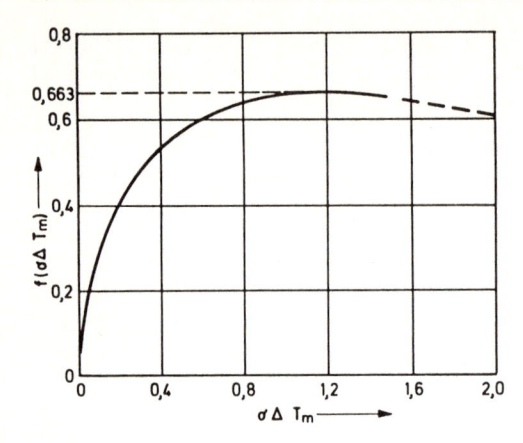

Fig. 1.3-11
Function for the calculation of the critical voltage

For further consideration, the above equation shall be written in the following form:

$$U = 2\sqrt{2}\sqrt{\frac{\lambda}{p_0\,\sigma\,e^{\sigma(T_u-T_0)}}} \cdot f(\sigma\Delta T_m) \quad \text{with} \quad \Delta T_m = T_m - T_u.$$

The function $f(\sigma\Delta T_m)$ is shown in Fig. 1.3-11.

A physically meaningful solution apparently requires that with increasing voltage U a higher value of maximum temperature T_m follows. But this condition is no longer satisfied in the region to the right of the maximum. The highest value 0.663 occurs at $\sigma\Delta T_m \approx 1.2$ and corresponds to the desired critical voltage U_k of the thermal breakdown. For that we obtain:

$$U_k = 1.875\sqrt{\frac{\lambda}{p_0\,\sigma\,e^{\sigma(T_u-T_0)}}} \quad \text{with} \quad p_0 = \omega\,\epsilon_0\,\epsilon_r\,\tan\delta_0.$$

In the commonly occurring case of only one-sided cooling of the plate, on integration from x = 0 to x = s half the value of U_k is obtained. Surprisingly, the critical voltage does not depend upon the plate thickness s but, for a given ambient temperature, only upon material properties. For the usual high-voltage insulating materials, values in the range of 50 kV to 500 kV are obtained at 50 Hz; with increasing ambient temperature U_k decreases rapidly.

For the example of an oil-paper dielectric at 50 Hz and 20 °C, the following values are approximately valid:

$\lambda = 0.002$ W/cm K ;
$\epsilon_r \tan\delta_0 = 0.016$;
$\sigma = 0.02$ K^{-1} .

For one-sided thermal conduction and an ambient temperature of 20 °C we have, using this data:

$U_k = 444$ kV .

For an ambient temperature of 100 °C it follows that

$U_k = 199$ kV .

1.3 Breakdown of solid insulating materials

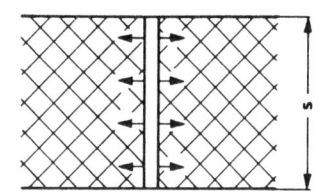

Fig. 1.3-12 Model for the calculation of the critical voltage of local thermal breakdown

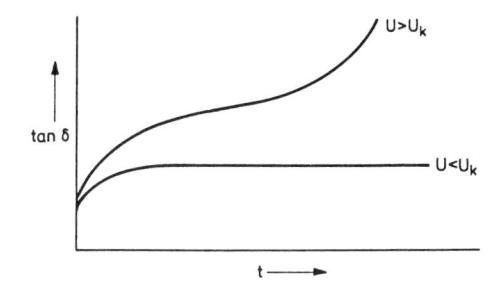

Fig. 1.3-13 Time dependence of the loss factor during tests for thermal stability

In the model with thermal conduction over the electrodes as in Fig. 1.3-10, the temperature distribution between electrode sections lying opposite each other is always the same. One can therefore speak of a "global thermal breakdown". In contrast, *K. W. Wagner* in his investigations according to Fig. 1.3-12 has assumed that a thin canal of increased conductivity exists in the dielectric and radial heat conduction takes place from it.

This model of a "local thermal breakdown" in the case of plates leads to an expression of the form

$$U_k \sim \sqrt{s} \;,$$

which in many cases also agrees very well with the experimental evidence [*Lesch* 1959].

Numerous theoretical investigations on the thermal breakdown of plate and cylinder configurations are known from the literature [*Dreyfus* 1924; *Berger* 1926; *Whitehead* 1951; *Franz* 1956].

The assumptions of the theory are only partly fulfilled in practical cases; hence the calculation yields only approximate values and cannot replace experimental verification of the thermal stability. This is done by holding the insulation concerned under voltage for a long time at operating conditions, then, after the thermal transients have died away, a static maximum temperature, i.e. a constant loss factor, sets in Fig. 1.3-13 shows the possible results of such a stability test which permits non-destructive determination of U_k. These tests are important for bushings, power capacitors, and cables among other things.

1.3.4 Partial discharge breakdown[1])

Partial discharges (PD) are electric discharges which bridge only part of the insulation; they are usually pulse shaped. This form of incomplete breakdown can take place in gas-filled cavities of a dielectric, or also at electrodes with small radius of curvature when these are not completely embedded in the solid insulating material. Partial discharges can, through long-period mechanisms, lead to a complete breakdown of the insulation, particularly during stressing with alternating voltages. The effective mechanisms are of a rather complex nature and therefore do not easily lend themselves to unified description. Despite extensive literature, comprehension of these mechanisms is still unsatisfactory. Nevertheless, in the following we shall attempt to describe briefly the mechanisms in a few technically rather important configurations.

---

[1]) Comprehensive treatment in [*Whitehead* 1951; *Kreuger* 1964; *Alston* 1968; *Bartnikas, McMahon* 1979].

Gas-filled cavities in a homogeneous insulating material, e.g. in plastics, should be considered as weak spots and are therefore basically undesirable. In laminated materials (e.g. hardboard) they are indeed unavoidable. Finally, cavities even occur in homogeneous dielectrics as a consequence of high electric stress. The formation of PD canals, known as "treeing", belongs to this category.

a) Configuration with internal partial discharges

According to Section 1.2.3a the incomplete breakdown in gases in a strongly inhomogeneous field is described as external partial discharge. The immediate transition into complete breakdown is impeded here by the fact that the leading electric field is only strong enough to sustain the discharge in the vicinity of the electrode with small radius of curvature. Partial discharges in or on solid insulating materials may also be considered as an incomplete breakdown in a gaseous dielectric. A few typical PD configurations are represented schematically in Fig. 1.3-14. The rapid extension into a complete breakdown is prevented here by the solid insulating material, namely by the finite dimensions of the gas volume available for the discharge and by the restriction of the discharge current. For not too small wall thicknesses the extension into complete breakdown occurs mainly through PD canals which are themselves gas-filled cavities.

Sensitive measuring methods allow the determination and evaluation of internal partial discharges, the occurrence of which is explained with the aid of various models [among others, IEC publication 270; *Kind* 1972 and the sources quoted therein].

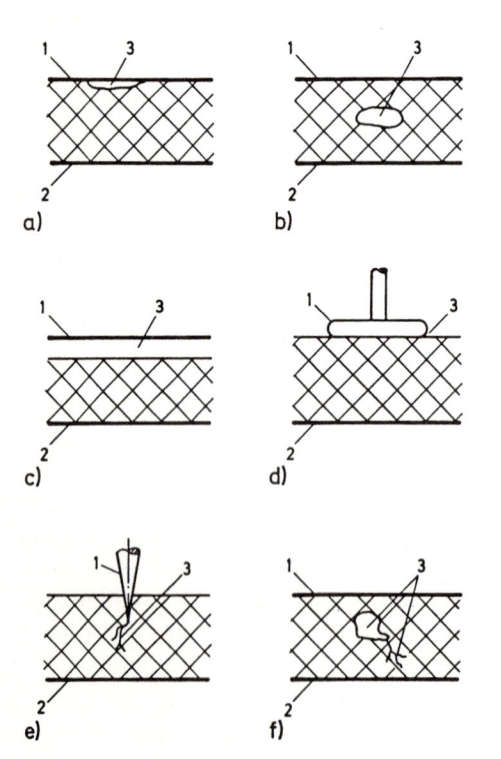

Fig. 1.3-14
Typical PD configurations
1, 2 electrodes, 3 partial discharge zone
a) cavity adjacent to electrode
b) cavity in insulator
c) detached electrode
d) electrode placed on surface
e) electrode point with PD canal
f) cavity with PD canal

1.3 Breakdown of solid insulating materials

b) Calculation of inception voltage

The field strength in the cavity can be calculated for configurations with known geometry, provided the rest of the dielectric behaves ideally and surface effects at the walls of the cavity are neglected. Fig. 1.3-15 shows the three simplest cases and states how large the constant field strength E_i in the cavity is for a homogeneous advancing field E; ϵ_r is the dielectric constant of the remaining dielectric [*Lautz* 1976]. In a similar manner E_i can be calculated for cavities in rotational ellipsoid form with arbitrary axis ratios [*Philippow* 1976].

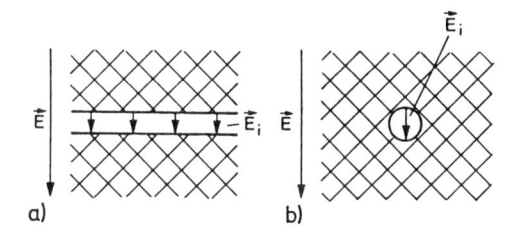

Fig. 1.3-15 Configurations for the calculation of the inception field strength
a) plane gap perpendicular to $\vec{E}$:
$$E_i = \epsilon_r E$$
b) cylinder perpendicular to $\vec{E}$:
$$E_i = \frac{2\epsilon_r}{1+\epsilon_r} E$$
sphere:
$$E_i = \frac{3\epsilon_r}{1+2\epsilon_r} E$$

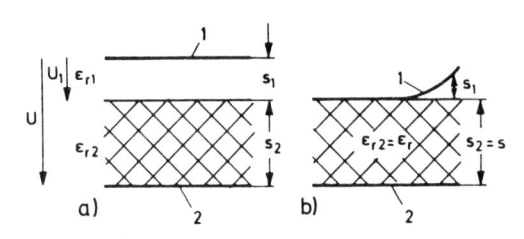

Fig. 1.3-16
Model for the calculation of inception voltage of mounted electrodes
a) arrangement with gap
b) arrangement with wedge

For the ignition of the discharge in the cavity validity of the Townsend mechanism as in Section 1.2.2a can be assumed. Since the field strength has the same magnitude throughout the cavity, yet the breakdown field strength E_d decreases with increasing gap width s, the breakdown should be expected at that point where the elongation in the direction of the field is greatest. The calculation procedure shall be demonstrated on the following example:

Let a spherical cavity in EP moulding ($\epsilon_r = 3$) with $s = 2$ mm diameter be filled with air at pressure $p = 0.5$ bar. According to Fig. A 2.1-1 for $ps = 1$ bar mm, the breakdown voltage $U_d = E_d s = 4$ kV. From $E_i = E_d$ it follows for the inception value of the leading field strength

$$E_e = \frac{1+2\epsilon_r}{3\epsilon_r} E_d = 1.6 \text{ kV/mm} .$$

A commonly occurring configuration in technical insulation systems is the insulating plate with the electrode mounted on it. For derivation of the inception voltage the configuration as in Fig. 1.3-14c is assumed in which there is a plane gap between the electrode and the insulating material [*Halleck* 1956]. If, according to Fig. 1.3-16a, the voltage at the gap is designated U_1 and U is the total voltage, we have:

$$U_1 = U \frac{s_1/\epsilon_{r1}}{s_1/\epsilon_{r1} + s_2/\epsilon_{r2}} .$$

U reaches the inception value U_e when $U_1 = U_d$, and we have:

$$U_e = U_d \left(1 + \frac{\epsilon_{r1}}{s_1} \cdot \frac{s_2}{\epsilon_{r2}}\right).$$

For gas-filled gaps we may put $\epsilon_{r1} = 1$; according to the Paschen law U_d depends upon ps_1. With $\epsilon_{r2} = \epsilon_r$ and $s_2 = s$ we have:

$$U_e = U_d(ps_1) \cdot \left(1 + \frac{1}{ps_1} \cdot \frac{ps}{\epsilon_r}\right).$$

Thus, for a certain value of ps/ϵ_r, U_e depends only on ps_1. If the electrode mounted as in Fig. 1.3-16b is now considered, it can be seen that the discharge begins in the wedge and that s_1 can be chosen so that U_e will be a minimum.

The results of numerical evaluation of this relation using the values of experimentally determined Paschen curves for air and SF_6 are represented in Fig. 1.3-17a whereby $\hat{U}_e$ should be understood as the peak value of the alternating inception voltage.

The comparison with the plotted measured values for air and SF_6 shows, in part, considerable deviation between measurement and calculation. This can be explained by the neglect of surface effects, by deformation of the insulating material in the region of the electrode edges and by other effects not taken into account in the model.

Another derivation, using a capacitive network as equivalent circuit, yields for the inception voltage

$$\hat{U}_e = \sqrt{2}K \left(\frac{ps}{\epsilon_r}\right)^\alpha.$$

For the exponent, in the case of small insulation thicknesses, a value of $\alpha = 0.5$ is obtained [*Philippow* 1956], which decreases with increasing thickness towards $\alpha = 0$ [*Rayes* 1978].

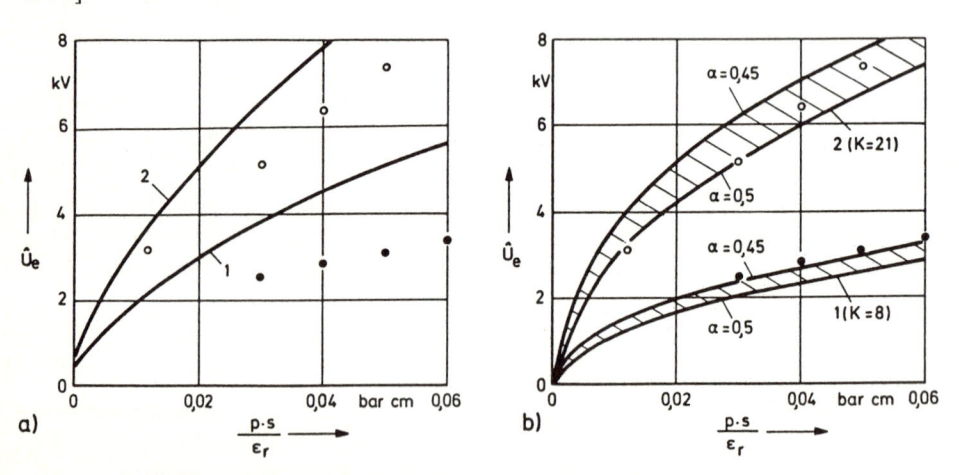

Fig. 1.3-17 Inception voltage for mounted electrodes in air (1) and SF_6 (2) Comparison of measurements and calculations
a) curves calculated using the Paschen law, measured values after *Brand* for PE in SF_6 (○), after *Kappeler* for hardboard in air (●)
b) curves calculated from $\hat{U}_e = \sqrt{2}K \left(\frac{ps}{\epsilon_r}\right)^\alpha$, measured values as in a)

1.3 Breakdown of solid insulating materials

The basic relationship between p, s and ϵ_r described here also corresponds, to a first approximation, to the result of experimental investigations [*Kappeler* 1949; *Kind* 1972; *Brand* 1973; *Rayes* 1978]. By the appropriate choice of the factor K the experimental data can be explained quite well most of the time.

For the usual range of

$\alpha = 0.45 ... 0.5$,

we have

for ambient medium air $K \approx 8$
for ambient medium SF_6 $K \approx 21$.

Here $\hat{U}_e$ is obtained in kV, if ps/ϵ_r is inserted in bar · cm.

Fig. 1.3-17b shows the regions overlapped by the approximation parabola with $\alpha = 0.45...0.5$ and in addition the same measured values as in Fig. 1.3-17a. For dimensioning the insulation a value of $\alpha = 0.5$ may be assumed since this results in lower values of $\hat{U}_e$.

Experience shows that the above relationship for the inception voltage is also valid for configurations with electrodes mounted on the dielectric and surrounded by an insulating liquid. This is quite in accordance with the knowledge of breakdown in liquids. For a metal or graphite edge the following is quoted for p = 1 bar [*Kappeler* 1949; *Böning* 1955]:

surrounding medium mineral oil $K \approx 30$.

To understand the inception of partial discharges in practical insulation systems, extensive investigations have been carried out on test samples with artificial sealed cavities [*Weniger* 1975; *Kübler* 1978]. It was noticed then, that, apart from the dimensions and shape of the cavities, it is above all their previous stress history which determines the inception voltage. Surface effects and variation in gas composition can cause time dependence of the ignition behaviour.

c) Mechanisms of partial discharge breakdown

During long-term electric stressing of an insulation, internal partial discharges can damage the dielectric as a consequence. This is equivalent to the prognosis that the ageing of insulating materials by means of partial discharges occurs above all during alternating voltage stress due to the periodic repetition of the ignition process; the following comments refer to this. Nevertheless, cavities with partial discharges can cause appreciable alteration of the electric strength for high direct voltages too; but this case is not a long-term mechanism [*Shihab* 1972].

The most important effects arising in these mechanisms are:

heating
erosion
chemical effect
charge carrier injection.

Partial discharges always result in additional dielectric losses, designated ionization losses. They occur in local concentration and so represent almost point-like heat sources. Estimations have shown that only in the high-frequency case (MHz range) can partial discharges cause inadmissible local heating and so lead to a thermal breakdown.

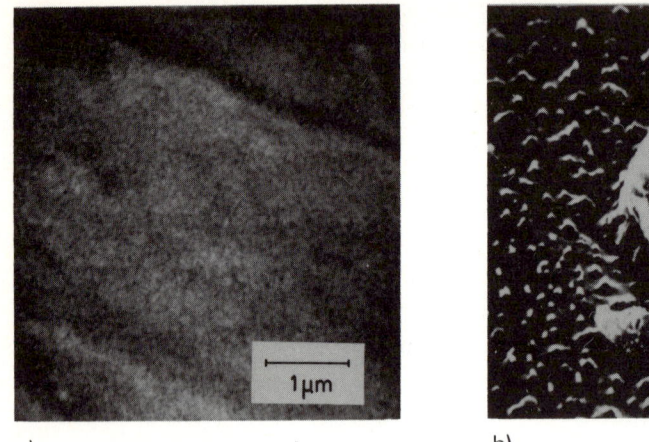

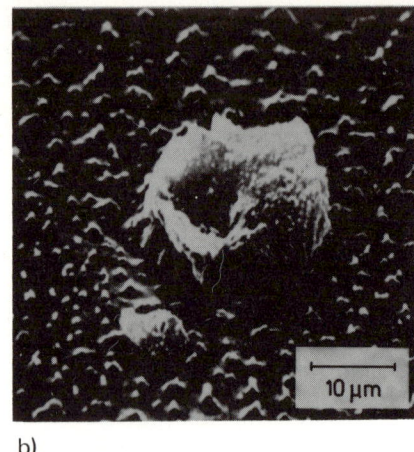

Fig. 1.3-18 Scanning electron micrographs of EP moulding surfaces
a) unstressed b) after 72 h PD stressing with 6 kV/mm

In gas filled cavities of a dielectric with partial discharges acceleration of electrons and ions occurs by means of Coulomb forces. The bombardment in particular of the walls of the insulating material by ions causes erosion, i.e. mechanical excavation of the material. On the other hand in air-filled cavities, seemingly crystalline deposits are the consequence of partial discharges; these can be made visible by scanning electron microscopy (Fig. 1.3-18) [*Salvage* et al. 1975]. The originally often very smooth surface can be roughened so yielding starting points for the partial discharge canals.

Life-time measurements on plastic test objects with artificial cavities have shown that for partial discharges too, an initial induction period should be expected if the surface of the insulating material exposed to the discharge is mechanically undamaged [*König* 1962; *Kodoll* 1974]. Where surfaces are mechanically prepared or for those sufficiently eroded by discharges, the destructive phase with the formation of canals sets in comparatively rapidly. On the other hand, PD breakdown as a consequence of the reduction, by uniform erosion, of the wall thickness of the remaining dielectric has not hitherto been observed.

Of great significance for the occurrence of a breakdown must indeed be chemical effects on the surfaces of an insulating material subjected to partial discharges. In oil-impregnated capacitors the X-wax formation at the edges of the metal foils, and in diphenylchloride impregnated capacitors the formation of hydrochloric acid, are known to be the cause of breakdown [*Liebscher, Held* 1968]. Even in foil insulation systems in gases, partial discharges can lead to breakdown by means of chemical effects [*Rao* 1968; *Schon* 1977]. In the case of plastic insulations with thicker walls, chemical effects are equally significant, particularly when moisture is also present. Basically all of the chemical reactions known from simple molecular chemistry are possible on the macromolecule as long as the appropriate environmental influences are present. For instance, at both high temperature and high humidity, epoxy resins are relatively easily hydrolysed [*Vogelmann, Henry* 1972]. Further, it is known that during electric discharge in air, besides other decomposition products, nitrous oxide (NO) is produced which together with moisture

1.4 Breakdown of liquid insulating materials

forms nitric acid (HNO$_3$) [*Schon* 1977]. Chemical changes in the gas-filled cavity and the solid material may also be promoted by the short-wave radiation occurring during ionisation processes.

Another ageing mechanism is the injection of charge carriers, preferentially electrons, into the insulating material. These penetrate the insulating material and are caught in traps, the result of which can be electron conduction. In PE foils it has been proved, using a configuration with air gap as in Fig. 1.3-14c, that penetration of insulating material by an electron wave can result in such high conductivity that the insulation becomes thermally unstable [*Boeck* 1967; *Rao* 1968]. In configurations as in Fig. 1.3-14d, on the other hand, starting from energy-rich surface discharges on the insulating material surface, electrons can penetrate the insulating material with the aid of a high space charge field strength and cause PD canals under the surface; these can on continued stressing develop into a complete "anomalous breakdown", of which Fig. 1.3-19 shows an example [*Dronsek* 1967].

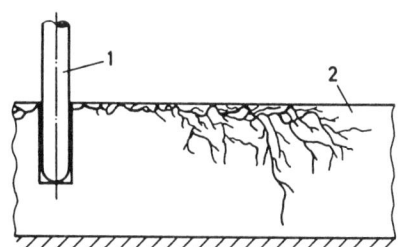

Fig. 1.3-19
Schematic representation of the origin of an anomalous breakdown:
1 high-voltage electrode
2 insulating plate with PD canals

1.4 Breakdown of liquid insulating materials[1])

In most applications insulating liquids, besides insulating voltage-carrying parts, have to satisfy further requirements. For example, they also serve to cool the windings and cores in transformers, to extinguish the arc in circuit breakers, or, as an impregnating medium in capacitors, they increase the dielectric constant of the paper dielectric.

The behaviour of technical insulating liquids in the electric field differs fundamentally from that of gases and solids. It is critically governed by impurities, by the ageing condition as well as by space charges. As a consequence of this there is no unified breakdown theory, even though certain mechanisms are beyond doubt. Only these shall be discussed briefly here, whereas the effect of refining and ageing on the breakdown performance, particularly critical in insulating oils, will be discussed in detail in Section 2.5.1.

1.4.1 Electric strength of technical configurations with insulating liquids

In high-voltage technology mineral oils are predominantly used as liquid insulating materials, namely for open insulating paths in conjunction with insulating supports, as well as for impregnation of laminated materials, especially soft paper and pressboard. Very often they are mineral oils of low viscosity, designated transformer oils; their viscosity, which is strongly temperature dependent, is matched to the application purpose (cooling, impregnation) by mixing with suitable distillates.

---

[1]) Comprehensive treatment in [*Strigel* 1955; *Imhof* 1957; *Kok* 1963; *Alston* 1968; *Gänger* 1981] and others.

As a result of their extraction from crude oil, insulating oils are a mixture of several hydrocarbons with different properties. In their application in large technical systems, as in transformers, and also in some cases by contact with the atmosphere, technical insulating oils also contain impurities in the form of dissolved gases, liquids (e.g. water, acids), as well as conducting and non-conducting particles (e.g. fibre pieces, sludge). Experience shows that the presence of these impurities determines the breakdown behaviour in practice much more than the properties of the ideally pure insulating liquid itself [Kok 1963]. In fact, only liquefied gases represent really pure insulating liquids; for very low temperatures liquid nitrogen (LN_2) or liquid helium may become a practicable proposition for future applications in low temperature technology.

As synthetic liquid insulators chlorinated diphenyls are used for the impregnation of the paper dielectric in power capacitors. Compared with mineral oils they have approximately twice the dielectric constant. A further advantage, the non-inflammability, was responsible in the early days for their application in indoor distribution transformers; today however, dry transformers with epoxy resin insulation are preferred for these applications.

The following table contains guiding values for some of the properties of insulating liquids:

	transformer oil	chlorinated diphenyls	liquid nitrogen
density in g/cm^3	0.9	1.4	0.8
ϵ_r	2.3	5.5	1.4

Besides depending upon the impurities, the electric strength also depends upon several other parameters, particularly upon pressure and the stress duration. During impulse voltage stressing the breakdown field strength of a configuration is many times the value for alternating voltages [Strigel 1955]; in a homogeneous field in practical insulations one may expect values up to $E_d = 200$ kV/cm. The impulse voltage-time curve of an electrode configuration in transformer oil reproduced in Fig. 1.4-1 gives an idea of the effect of stress duration [Kratzenstein 1969].

Typical for breakdown measurements in insulating liquids with impurities is large dispersion and the occurrence of irregular pre-discharges, even in a homogeneous field. Moreover, in pure liquid gaps breakdowns can occur with subsequent self-healing.

Fig. 1.4-2 shows the result of measurements of the breakdown field strength E_d and the dissipation factor $\tan\delta$ at 50 Hz as a function of water content v [Holle 1967]. The reduction of E_d on exceeding $v = 50 \cdot 10^{-6}$ can be attributed to the fact that a transition from solution to emulsion takes place. For a breakdown field strength of at least 200 kV/cm a residual water content of $v < 10^{-5}$ must be ensured.

In contrast to dissolved water vapour, dissolved gases have in general no effect upon the electric strength of insulating liquids, apart from the ageing processes due to oxygen. However, the condition of supersaturation could be critical if beyond the equilibrium condition dissolved gases appear in the form of tiny bubbles released by mechanical

1.4 Breakdown of liquid insulating materials

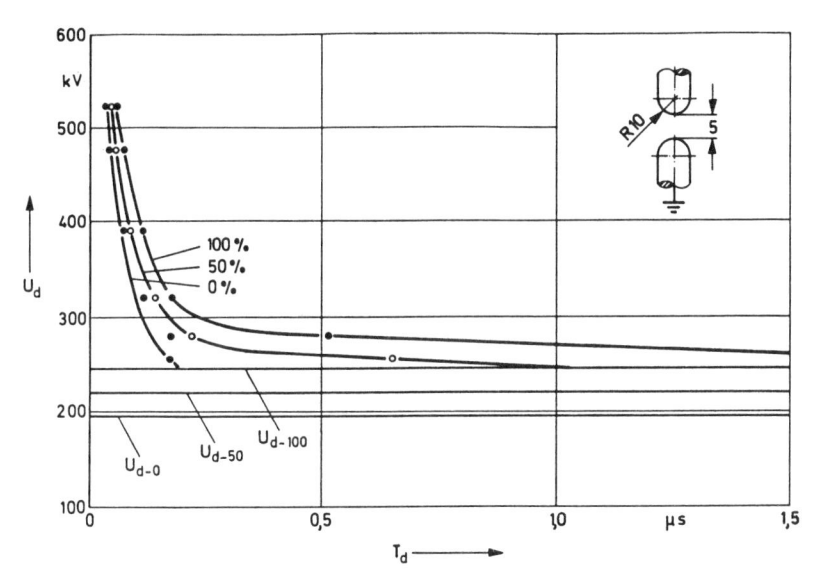

Fig. 1.4-1 Impulse voltage-time band of the rod-rod electrode configuration in oil for negative impulse voltage

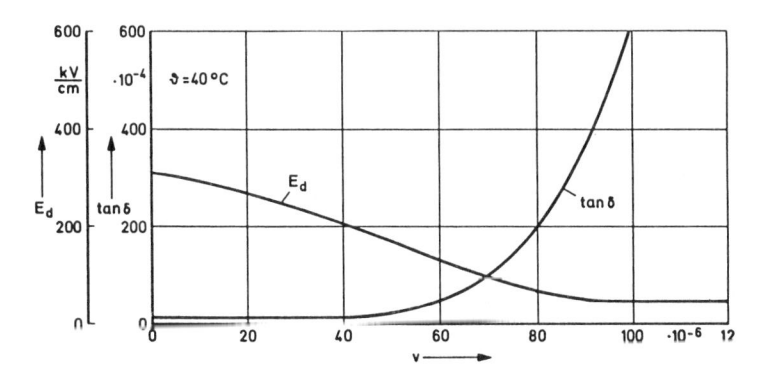

Fig. 1.4-2 Breakdown field strength and dissipation factor of transformer oil as a function of water content (mass fraction)

vibrations (forced circulation cooling) or by high electric field strengths [*Kind* 1959; *Strigel, Winkelnkemper* 1961].

Liquid insulating materials are used as impregnants in the dielectric of capacitors, in the soft paper and pressboard insulation of transformers, as well as in oil-impregnated paper cables. Here very high electric strengths are reached but at the cost of effective convection cooling. The following table gives a few guiding values for physical properties at 20 °C:

	oil/paper	oil/pressboard	chlorinated diphenyls/paper
density in g/cm^3	1.1	1.2	1.5
ϵ_r	3.6	4.5	5.5

The electric strength of these mixed dielectrics is so high that continuous operating field strengths of 100 kV/cm and above can be permitted. For short-duration stressing values of E_d up to 1 MV/cm can be measured. Especially at high ambient temperatures however, the possibility of a thermal breakdown as described in Section 1.3.3 must be considered.

In the special case of extremely short-duration stressing, even water shows a high breakdown field strength. Depending upon the experimental conditions, for spacings of a few mm one can obtain values between 100 kV/cm and 500 kV/cm, whereby the stress duration may not exceed a few 10 µs [*Kuzhekin* 1972; *Steudle* 1974]. As in technical insulating liquids the electric strength increases greatly with pressure. With these properties, and in combination with its high dielectric constant of $\epsilon_r \simeq 80$, water is very well-suited as an impregnating medium for configurations exposed to the risk of flashover; this property is occasionally made use of in high-voltage impulse setups [*Dokopoulos, Steudle* 1972].

1.4.2 Breakdown mechanisms

The breakdown of liquids cannot be described by a single unified theory and there is also strong dependence of the observed phenomena on the technical boundary conditions. Hence, only the two most important types of breakdown will be discussed here.

a) Intrinsic breakdown

As in gases and solid insulating materials, an avalanche breakdown is also possible in very pure liquids. If the ionization condition according to Section 1.2.1

$$E \lambda \geq U_i$$

is assumed, reduction of the mean free path λ due to the liquefaction of the gas should be compensated by a corresponding increase in the ionization field strength. An estimate for the example LN$_2$ shows that this theoretically desirable value of the field strength cannot ever be achieved in an experiment. Rather, one should expect the kinetic energy of the electrons, even for the experimentally determined values of the field strength, to be sufficient to effect partial vaporisation of the liquid by way of collision processes with the liquid molecules. In the small gas bubbles so formed, on account of the larger mean free path λ, the prevailing field strength can initiate collision ionization and avalanche formation. This model, in conformity with the measured results for LN$_2$, results in breakdown field strengths of about 300 kV/cm [*Peier* 1976].

In technical insulating oils more complicated mechanisms may be assumed as a consequence of the presence of different components. In an electrode configuration with an insulating liquid in a d.c. field, a current density S appears which only reaches a nearly constant value after a few minutes (Fig. 1.4-3a). The cause of this phenomenon is the

1.4 Breakdown of liquid insulating materials

Fig. 1.4-4 Formation of a fibre bridge

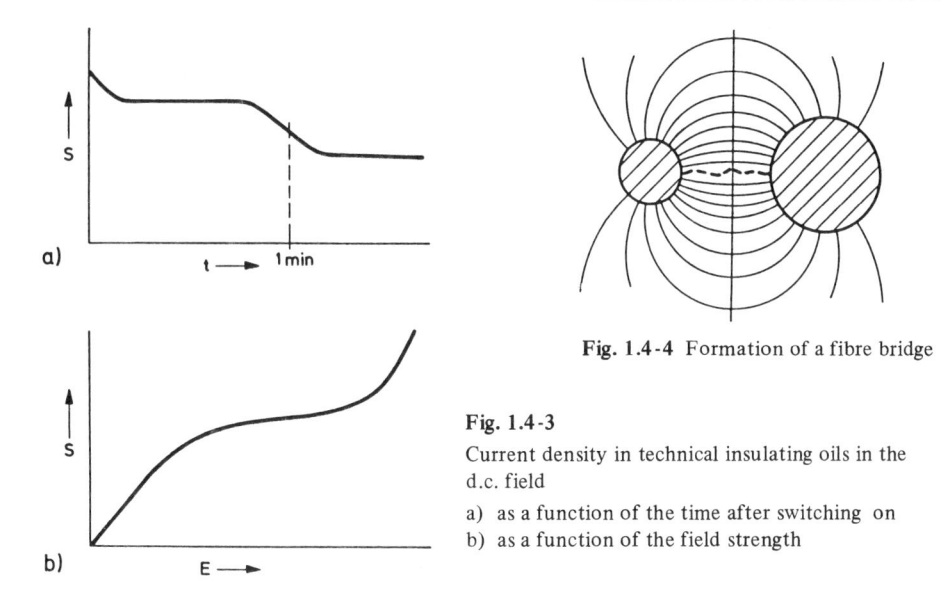

Fig. 1.4-3
Current density in technical insulating oils in the d.c. field
a) as a function of the time after switching on
b) as a function of the field strength

presence of charge carriers of different mobility. Finally, the steady current is determined by the heavy electrolytic ions which are formed by dissociation [*Whitehead* 1935]. For low field strengths Ohm's law is approximately valid, until a saturation current sets in, as in gases (Fig. 1.4-3b). If the field strength E is increased further, the current increases disproportionately until breakdown occurs. According to the expression for the current density

$$\vec{S} = qnb\vec{E} \quad (q = \text{ionic charge}),$$

an increase in the charge carrier density n may be expected from this since there is no reason for the mobility b to change. The presumption that, similar to a breakdown in liquefied gases, the charge carrier multiplication may be attributed to collision ionization in the gaseous or vapour part of the liquid, leads to the description of this breakdown as a "masked gaseous breakdown". This model also explains the experimentally observed increase in the electric strength of insulating liquids with pressure.

b) Breakdown by fibre bridge formation

Technical insulating liquids always contain macroscopic impurities in the form of fibrous particles of cellulose, cotton or other materials. Especially when these particles have absorbed moisture from the insulating liquid, forces act upon them moving them into the zone of higher field strength and also aligning them in the direction of the field [*Kind* 1972]. In contrast to the fibre particles, gas bubbles, because of their lower dielectric constant, are removed from the region of highest field strength.

In this manner, as indicated in Fig. 1.4-4, a fibre bridge between the electrodes can result, and this represents a conducting canal. The resistance loss can result in vaporization of the moisture contained in the particles with subsequent gas breakdown. This phenomenon can also be interpreted as a local thermal breakdown originating from a weakly conducting canal. The technically extremely important occurrence of a breakdown by the formation of fibre bridges prohibits high electric stressing of free oil gaps. Fibre bridge breakdown can be effectively prevented by insulating screens which should be arranged perpendicular to the electric field if possible. A further effective measure is to embed the electrodes in a solid insulating material, preferably using a paper bandage.

1.5 Breakdown in high vacuum[1])

All breakdown theories for gaseous, liquid and solid insulating materials assume that the insulating material is made conducting by ionizing processes. In high vacuum ($p \leqslant 10^{-5}$ mbar), the mean free paths are so large that collision processes in the rest of the gas become meaningless for the breakdown mechanism. Rather, the mechanisms at the electrodes are critical to the breakdown behaviour.

If direct or alternating voltage is applied to a vacuum gap, preliminary currents start very much below the breakdown voltage and increase exponentially with increasing voltage. It has been shown on point-plate gaps [*Dyke* et al. 1953], that the preliminary currents follow the Fowler-Nordheim equation for field emission [*Finkelnburg* 1967]:

$$S = 1.55 \cdot 10^{-6} \frac{E^2}{W_a} \exp\left[-\frac{6.9 \cdot 10^7 \cdot W_a^{3/2}}{E}\right].$$

Here S is the current density in A/cm², E is the field strength in V/cm and W_a is the work function in eV.

For large area electrodes field emission currents can be measured which are larger by several powers of ten than the preliminary currents expected according to the Fowler-Nordheim relation. This is caused by micropeaks on the surface of the electrode, which locally enhance the electric field.

Many breakdown hypotheses have been developed to explain the mechanisms in vacuum gaps. The cathodic breakdown hypothesis [*Alpert* et al. 1964] assumes that the field emission current, at a micropeak on the cathode, above a critical current density leads to so much heating that the micropeak evaporates explosively. In the metal vapour then formed ionizing collision processes take place. If sufficient charge carrier multiplication is achieved breakdown of the vacuum gap occurs across the ionized metal vapour cloud.

In the anodic breakdown hypothesis it is assumed that the electrons released from the cathode by field emission and accelerated in the electric field to an energy W = eU, heat up the anode to such an extent that the anode material vaporizes. The metal vapour is ionized by collision processes and reinforces the electron emission back at the cathode. For a sufficiently high vaporization rate at the anode gas breakdown occurs within the metal vapour cloud [*Maitland* 1962]. The theory described is supported by the observation that for breakdown the pressure in the experimental tank increases by about a factor of ten and anode material can be found on the surface of the cathode [*Schmidt* 1979].

According to the clump hypothesis [*Cranberg* 1952], vacuum breakdown is initiated by free metal particles residing on the electrodes which are torn and accelerated by field forces and vaporize on collision with the opposite electrode.

---

[1]) Comprehensive treatment in [*Alston* 1968; *Meek, Craggs* 1978; *Latham* 1981] and others.

1.5 Breakdown in high vacuum

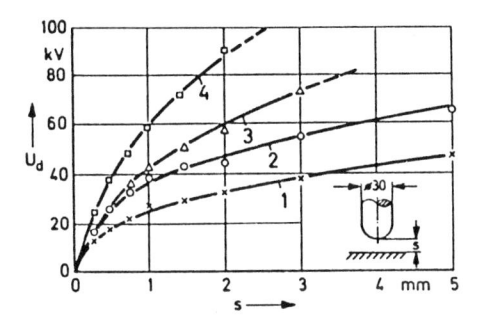

Fig. 1.5-1 Effect of electrode material on the breakdown voltage of a high-vacuum configuration with positive and negative direct voltage in a weakly inhomogeneous field

1 zinc   2 aluminium   3 copper   4 steel

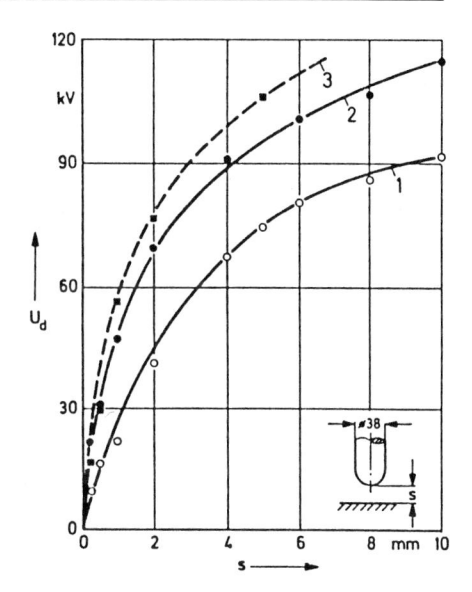

Fig. 1.5-2 Effect of electrode material and electrode temperature on the breakdown voltage of a high-vacuum configuration with alternating voltage

1 copper electrodes T = 293 K
2 copper electrode, plate T = 80 K
3 steel electrodes T = 293 K

The electric strength of a vacuum gap of spacing s in the homogeneous or weakly inhomogeneous field for direct voltage stressing satisfies the relation $U_d \sim \sqrt{s}$ [*Zeibig* 1966]. This relationship is confirmed for impulse voltage, for which, in the region of smaller breakdown times ($t_d < 0.1\,\mu s$), a steep increase of the impulse voltage-time curve is observed [*Bauer* 1971].

As Fig. 1.5-1 shows [*Schmidt* 1979], the electrode material also influences the electric strength: under otherwise identical conditions, the d.c. breakdown voltage increases with higher melting point of the electrode material. This behaviour conforms with the breakdown hypotheses already mentioned. Cooling the electrodes has the same effect as a higher melting temperature and increases the electric strength of the configuration (Fig. 1.5-2).

Breakdowns under direct voltage lead to strong erosion of the anode. In contrast, the surface finish of the cathode is improved. From Fig. 1.5-3, which is a comparison of the sectional magnification of a cathode and an anode surface, one can recognize that even a single breakdown is followed by appreciable erosion of the anode in consequence [*Dohnal* 1981]. During a.c. stressing of homogeneous configurations both the electrodes are eroded equally because the electrodes act alternately as anode and cathode. Here, surface structures comparable with the anode surface shown in Fig. 1.5-3 are obtained. Due to the erosion of both the electrodes the breakdown voltage is lower than that for d.c. stress [*Schmidt* 1979].

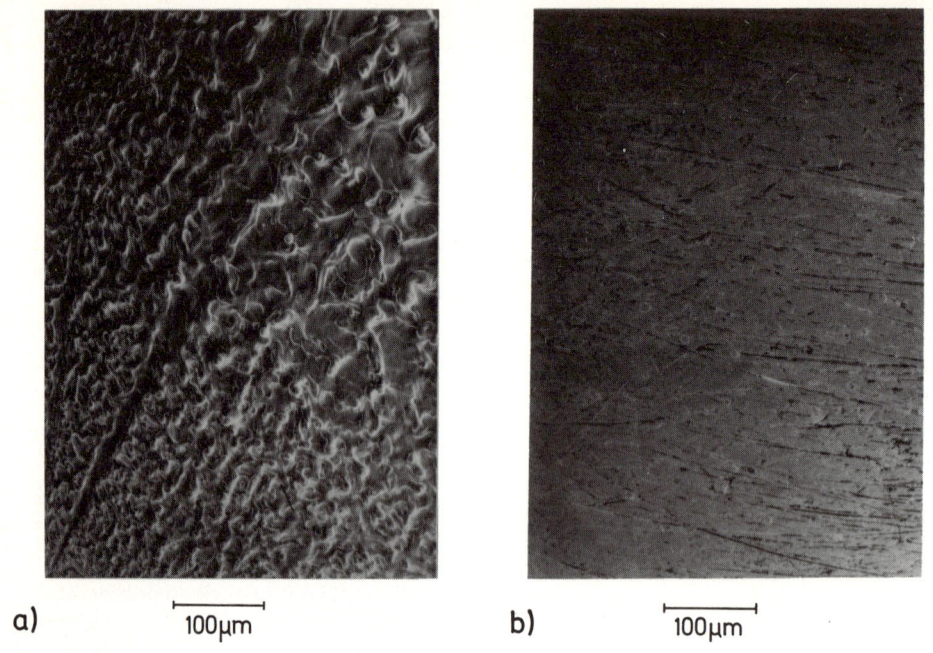

Fig. 1.5-3 Electrode surfaces after a vacuum breakdown
a) anode   b) cathode

In inhomogeneous configurations, on the other hand, the breakdown voltages under a.c. and d.c. stress are equal, since the alternating voltage breakdown occurs preferentially for negative polarity of the electrode with smaller radius of curvature and so distinct erosion can be observed only at the electrode with the larger radius of curvature. Thus comparable surfaces and breakdown voltages are obtained for both types of voltage.

1.6 Pollution flashover[1])

High-voltage setups often contain insulating bodies in a gaseous environment, which are stressed by a flashover. If a contamination layer develops on the surface of such an insulator, its electric strength can be enormously reduced. Above all, this is valid for insulators of overhead transmission lines or in outdoor switching stations, the long-term behaviour of which under atmospheric pollution is of great significance to the operating security of high-voltage networks. This chapter will deal with some fundamentals for this in particular.

1.6.1 Development and effect of contamination layers

At the boundary surfaces between solid and gaseous substances different physical mechanisms such as condensation and adsorption take place. Further, for insulators in atmospheric air one should expect contamination layers to form by accumulation of dust

---

[1]) Comprehensive treatment in [*Philippow* 1966; *Lambeth* 1972; *Holte* et al. 1979] and others.

1.6 Pollution flashover

particles. Since a general quantitative description of these processes is not possible, conclusions concerning the electrical behaviour of technical insulators shall be drawn here from investigations of simple models.

a) Moisture layer

Atmospheric air always contains appreciable quantities of moisture; thus the standard conditions prescribed by IEC for tests in air with 11 g water per m^3 at 20 °C and 1013 mbar mean a relative humidity of about 65%. Deposition of water molecules occurs on the insulator surface, which however, for clean experimental conditions, only results in a decrease of the flashover voltage U_d at relative humidity values $F > 50\%$ (Fig. 1.6-1) [*Link* 1975]. The reason for this behaviour is the formation of a mono-molecular film of water on the surface of the insulating material, even below the saturation humidity of the surrounding air. The reduction of the flashover voltage caused in this way depends on the form and surface finish of the insulator, as well as on the voltage shape, among other things.

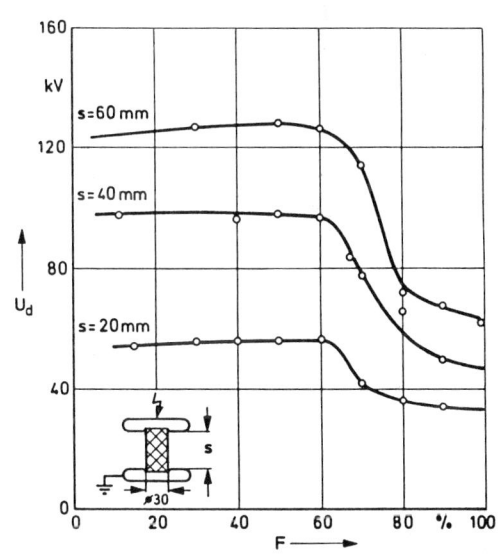

Fig. 1.6-1
a.c. flashover voltage of a clean epoxy resin support insulator in humid air at 30 °C
[*Link* 1975]

On outdoor insulators a cohesive water layer can be produced by rain, fog or dew on the undercooled insulator, especially in the early morning hours. The effect of moisture on the flashover voltage of insulators is also important for SF$_6$ insulated setups. The insulating gas used must be sufficiently dry to guarantee that the dew point is not reached at solid insulating material surfaces [*König* et al. 1977].

b) Conducting contamination layers

Dust-like impurities settle on the surface of objects in the atmosphere. As long as the thickness of the contamination layer so formed stays within limits and remains dry, this mechanism has just as little effect on the flashover voltage of the insulator as has a certain amount of humidity. However, if the contamination layer and moisture combine, a conducting contamination layer is formed which can cause an appreciable reduction in the electric strength of an insulator. For orientation purposes, a few guiding values for the flashover voltage of insulators with a highly conducting contamination layer, referred to their dry condition, are compiled below for different types of voltage [*Reverey, Verma* 1970]:

	reduction to about
lightning impulse voltage	90%
switching impulse voltage	50%
alternating voltage	20%
direct voltage	15%

It follows from this that the behaviour of insulators with contamination is of special significance to the operating stress under alternating or direct voltages.

In practice three basically different types of pollution commonly occur:

1. "Salt fog pollution" mainly occurs near the coast where salt fog can be carried several kilometres inland. Similar conditions can also prevail in the vicinity of roads strewn with salt in winter.
2. "Industrial pollution" should be expected in the neighbourhood of installations with corresponding emission, as in the case of cement factories, coal-firing installations or chemical industries. The deposit is mostly inert dust and salts mixed with dust which initially are dry and later, combined with moisture, cause electrolytic conductivity. Occasionally conducting contamination layers are formed by the solution of acid-forming gases such as SO_2 in water.
3. A special case is "desert pollution" which arises in desert regions by the deposition of dust. Wetting occurs by the early morning dew. Dust is transported by the wind so that deposition of the contamination layer occurs particularly on aerodynamically unfavourable locations, for example at the back sections of insulator sheds. For this type of contamination constructions with protected creepage path through the petticoats are unsuitable, since after some time these are completely filled up with dust. Long-rod insulators with horizontal, aerodynamically favourable shed designs have proved successful under these circumstances.

In all types of pollution a conducting contamination layer appears on the insulator surface in which a leakage current I flows on the application of a voltage U. Under the assumption of constant layer conductivity σ_s for axisymmetrical insulators according to Fig. 1.6-2, we have for calculation of the differential contamination layer resistance dR over the section dx:

$$dR = \frac{dx}{\pi D} \frac{1}{\sigma_s}.$$

Fig. 1.6-2
For the calculation of the leakage resistance of an axisymmetrical insulator with conducting contamination layer

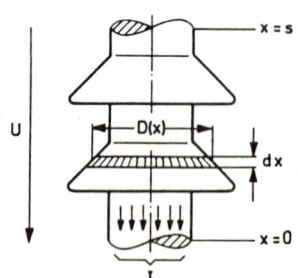

1.6 Pollution flashover

Integration over the creepage path length s_k gives the leakage resistance:

$$R = \frac{1}{\sigma_s}\frac{1}{\pi}\int_0^{s_k}\frac{dx}{D(x)} = \frac{1}{\sigma_s}K_f .$$

The form factor K_f can be determined by graphical integration, for example, from the shape of the insulator; for various designs it lies in the range 10 to 30.

Extensive investigations with alternating voltage on insulators of different construction with artificial and natural pollution, have proved that the layer conductivity σ_s is a useful measure of the degree of pollution. Fig. 1.6-3 shows that the a.c. withstand voltage U_{d-0} of the insulator decreases monotonically with increasing σ_s [*Nasser* 1962]. Instead of σ_s, the degree of pollution can also be expressed by the mass of salt coating per unit area, or by the salt content of the fog.

Fig. 1.6-3
a.c. withstand voltage U_{d-0} of a 110 kV long-rod insulator as a function of the layer conductivity σ_s
degree of pollution:
1 light 2 medium 3 heavy 4 very heavy

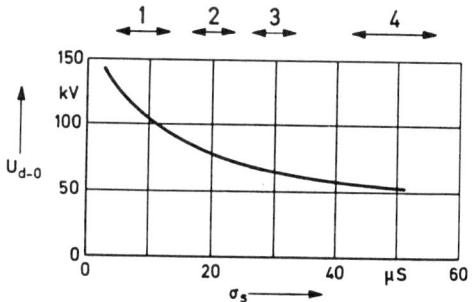

Under direct voltage the formation of a contamination layer is greatly influenced by the electrostatic forces acting on the dust particles. The electric field can lead to a very irregular pollution and so to inhomogeneity of the layer conductivity. The consequence is an even greater reduction in the flashover voltage with pollution than for alternating voltage.

1.6.2 Mechanism of pollution flashover

The phenomena leading to the flashover of a polluted insulator are extremely complicated and are indeed to a great extent governed by chance. Nevertheless, a highly simplified model can help to describe the most important processes.

a) Formation of dry bands

Fig. 1.6-4a shows the plane model of a configuration with a homogeneous contamination layer of conductivity σ_s [*Lambeth* 1971]. A leakage current flows which produces a linear potential distribution. This leads to a certain amount of drying of the contamination layer which, as in Fig. 1.6-4b, may possibly occur in bands; corresponding to the locally widely differing conductivities the potential then drops essentially at the dry band and the current temporarily becomes very small. Finally, as shown in Fig. 1.6-4c, bridging of individual dry bands by means of partial arcs occurs which ultimately can turn into a complete flashover. This latter is prevented if complete drying of the surface occurs, when once more a linear potential distribution results but with current values much lower than the danger level. The entire mechanism can be regarded as a race between uniform drying and the cascading of partial arcs.

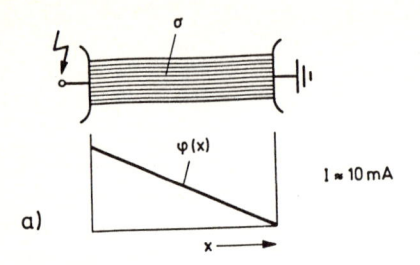

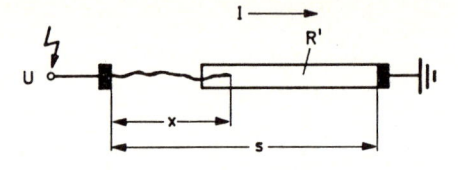

Fig. 1.6-5 Model of an insulator with dry band and contamination layer for determination of the critical values of leakage current I and voltage U

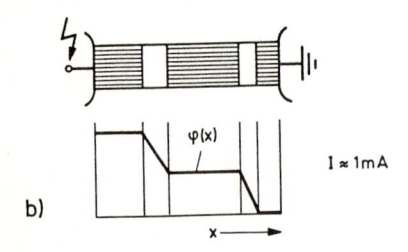

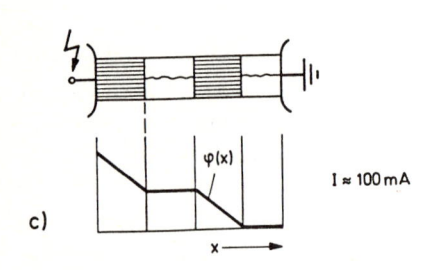

Fig. 1.6-4
Potential distribution $\varphi(x)$ in a plane contamination layer model
a) homogeneous conducting contamination layer
b) origin of dry bands
c) partial arcs

b) Stability considerations using the contamination model

A basic idea shall be developed using a model conceived by *F. Obenaus* in 1958. According to Fig. 1.6-5 the insulator with dry band and the conducting contamination layer is represented by the series connection of an arc path of length x and a resistance with a homogeneous resistive layer per unit length $R' = R'(I)$. With this model one may investigate whether, once ignited an arc protracts or extinguishes.

The total voltage U comprises the partial voltages across the arc and the contamination layer. With the arc field strength $E_b = E_b(I)$ we have:

$$U = E_b x + I R'(s - x) .$$

As the condition for extinction of the discharge it will be assumed that the voltage required by the arc for an extension on the basis of its characteristic $E_b(I)$ increases more rapidly than that available from the supply across the layer resistance:

$$\frac{\partial(E_b x)}{\partial x} > \frac{\partial(U - I R'(s - x))}{\partial x} .$$

If it is further assumed that the arc voltage forms only a small portion of U, then I becomes independent of x and as a condition for extinction we have the following:

$$E_b > I R' .$$

1.6 Pollution flashover

It then follows that forward growth of the arc must be anticipated if a certain critical current value

$$I_k = \frac{E_b}{R'}$$

is exceeded. A flashover criterion can be derived from this if the current dependence of the arc field strength and of the resistive layer are known. Fig. 1.6-6 shows a qualitative diagram of the boundary of the current regions for forward growth and extinction of the arc; it is assumed here that E_b as well as R' decrease with increasing current. In the hatched region the arc develops into a complete flashover.

For an approximate calculation of I_k we assume:

$$E_b = b \cdot I^{-n}$$
$$R' \approx \text{const.}$$

and finally obtain:

$$I_k = \left(\frac{b}{R'}\right)^{\frac{1}{n+1}} \sim \sigma_s^{\frac{1}{n+1}} \ .$$

It follows that a flashover can be expected if a certain maximum leakage current is present. In fact, it has been shown that, for the failure of an insulator with contamination layer, the peak value of the leakage current $\hat{I}_{max}$ immediately before a flashover is a characteristic parameter largely independent of the type of pollution [*Verma* 1976]. Fig. 1.6-7 shows the development of the leakage current up to complete flashover on the basis of a schematic representation of its time characteristic in the phases of Fig. 1.6-4.

More important in practice, however, is the critical voltage U_k corresponding to I_k. It can be derived from the model if one assumes that for the arc length in the critical region $x \ll s$ holds. Then we have:

$$U_k \approx I_k R' s = b^{\frac{1}{n+1}} R'^{\frac{n}{n+1}} s \sim \sigma_s^{-\frac{n}{n+1}} s \ .$$

The linear dependence of the flashover voltage upon the insulator length s has been proved experimentally to very high voltages. For the exponent $n/n+1$, values in the range

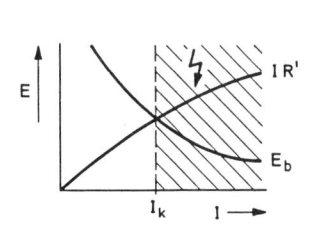

Fig. 1.6-6 Regions for extinction and forward growth of the arc in a contamination layer model as in Fig. 1.6-5

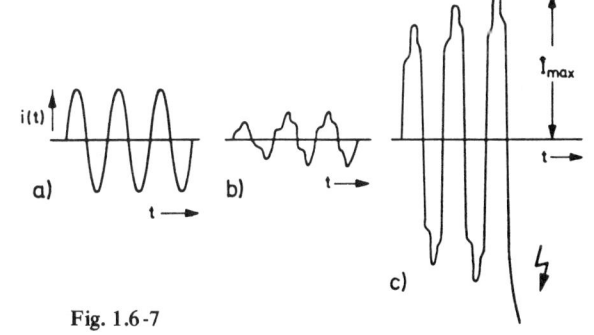

Fig. 1.6-7

Qualitative time curve i(t) of the leakage current in the phases of Fig. 1.6-4 for a complete flashover

0.2...0.6 were obtained [*Lambeth* 1971]. For a first rough estimate the simple assumption of n = 1 is adequate. It then follows that:

$$I_k \sim \sqrt{\sigma_s}$$

$$U_k \sim \frac{s}{\sqrt{\sigma_s}} \quad .$$

Even though complicated models may do better justice to the real conditions, one should not assume that their application can replace experimental testing and practical confirmation.

1.6.3 Pollution tests

Testing insulators for their contamination behaviour must take the physical mechanisms during development of flashover into particular account. Since this involves great expense and effort a few particulars shall be given here.

Deposition of the contamination layer on the insulator can be done either before beginning the voltage test or during the test. In the case of natural pollution, formation of the contamination layer depends on whether the voltage was on or not during increasing pollution. Because of the effect of the electric field on the thickness and distribution of the contamination layer, the results can vary considerably. Natural pollution has the advantage that the site conditions are accurately realized, but too much time is required and the reproducibility poor. In artificial pollution the contamination layer is deposited before or during the test, where the requirement is to realize as closely as possible the natural pollution condition at the future site.

Contrary to other high-voltage tests the pollution test may not be conducted as a short-duration test with increased voltage, since otherwise the contest between drying the contamination layer and cascading of partial arcs, essential for passing the test, would be interfered with. The test must therefore be performed at constant test voltage whilst the degree of pollution is increased from test to test until the sample fails.

Based upon the types of pollution described in Section 1.6-1, one distinguishes between two test methods[1]):

1. Salt fog test method
   At the beginning of the test the insulator is clean; the flowing contamination layer is formed during testing by deposition from the salt fog contained in the environment. This method is meant to simulate the effect of salt fog pollution and requires an expensive testing chamber, or at least a testing tent. The degree of pollution is expressed by the salt content of the fog.

2. Kieselguhr test method
   Before beginning the test an adhesive contamination layer is applied to the insulator, e.g. a mixture of kieselguhr and salt (pre-deposited pollution method). The necessary wetting either occurs during testing from the surroundings (clean fog method) or is already inside the layer from the start (flow-on method); in the latter case the test

---

[1]) For further details, see [IEC Publication 507].

1.6 Pollution flashover

can be performed in a normal testing hall. This method corresponds to the origin of industrial pollution; the degree of pollution is expressed either by the quantity of salt deposited or by the layer conductivity.

During pollution tests special demands are made on the test voltage sources too [*Kolossa* 1971]. The leakage current of an insulator with contamination layer consists, for alternating voltage, of a highly distorted alternating current, whose peak values reach the value of 1 A. But the test voltage may only drop minimally in consequence; the voltage source must therefore be comparatively stiff. Inadmissibly high transient voltage drops during direct voltage testing can often only be avoided by dynamic readjustment of the test voltage, which simulates a sufficiently low internal resistance of the test voltage source.

2 Insulating Materials in High-Voltage Technology

Dimensioning an insulation system requires exact knowledge of the type, magnitude and duration of the electric stress while simultaneously considering the ambient conditions. But, on the other hand, properties of the insulating materials in question must also be known, so that in addition to the proper material, the optimum, e.g. the most economical, design of the insulation system can be chosen. A particular problem in this respect is that the determination of the properties of insulating materials is done with model samples under standardized conditions, so that extrapolation of these model values to real insulation systems is often not unconditionally possible. Added to this, many values of the properties of the insulating materials are subject to considerable statistical scatter so that dimensioning an insulation system must be done with appropriate safety margins.

2.1 Requirements for insulating materials

The most important function of the insulating material is to insulate voltage-carrying conductors against one another as well as against earth. But, in addition, they must frequently perform mechanical functions and must be in a position to withstand certain thermal and chemical stresses. Such stresses very often occur simultaneously, so that the mutual effects of the various parameters must be known. Decisive for the economical application of an insulating material is, ultimately, its long-term or ageing durability under various types of stresses encountered in practice.

Depending on the weight given to the type of application, the following requirements are specified for the electrical properties of the insulating material:

    high electric strength, to realize small dimensions and low cost using as little material as possible,

    low dielectric losses, to keep the heating-up of the insulating material within limits,

    high tracking strength during surface stress, to prevent erosion or tracking,

    appropriate dielectric constant.

The mechanical requirements are a consequence of the fact that most of the insulating materials are simultaneously construction materials with load-bearing properties. Depending on the application, important properties are the tensile strength (e.g. overhead line insulators), the bending strength (post insulators in substations), the pressure strength (pedestal insulators of antennae) or the bursting-pressure withstand strength (circuit breaker insulators stressed by internal pressure). Mechanical characteristics such as modulus of elasticity, hardness, impact resilience etc. are significant in connection with the stress and the appropriate design.

Electrical equipment and setups are often exposed to increased temperatures during normal operation as well as during fault conditions. Specification of thermal properties, such as high thermal withstand strength, good shape-retention under heat, high thermal conductivity, low thermal expansion coefficient, non-inflammability, good arc-withstand strength etc. is the consequence.

Finally, an insulating material must also be insensitive to the ambient conditions at its site of application. Durability in oil, ozone resistance, impermeability, hygroscopic resilience, low water absorption, radiation stability can become necessary as additional requirements.

Technological properties such as good processability and workability, homogeneity, dimensional stability etc., which are important for economical production, should also be taken into account.

Insulating materials applied in technical high-voltage systems must, therefore, often satisfy a number of requirements some of which may even be contradictory. Thus the choice of an insulating material for a particular purpose often makes it necessary to find a compromise between the diverse requirements and the properties which satisfy them only incompletely.

2.2 Properties and testing of insulating materials[1])

2.2.1 Electrical properties

a) Breakdown field strength

The breakdown field strength is an extraordinarily important material property for dimensioning an insulation system, although it does not represent a constant specific to the material. What is more, it depends more or less strongly upon several influencing parameters as radius of curvature and surface finish of the electrodes, layer thickness, type of voltage, stress duration, pressure, temperature, frequency and humidity. For certain insulating materials and electrode configurations tabulated values of the strength are available (for instance, for air and SF_6 at standard conditions and in different configurations (see Appendix 2)). In other cases the breakdown strength of an insulation for a particular application needs to be determined experimentally each time.

For solid insulating materials, certain criteria are available from measurement of the breakdown voltage or the breakdown field strength on plate-like samples in a homogeneous or weakly inhomogeneous field. Gaseous and liquid insulating materials are tested between spherical segments[2]).

Fig. 2.2-1 shows, as an example, a standardized testing arrangement for the determination of the breakdown field strength of plates or foils up to sample thicknesses of 3 mm. In order to prevent gliding discharges along the surface of the insulating plate, the entire arrangement is embedded in an insulating liquid with an appropriate dielectric constant. An electrode arrangement of spherical segments is shown in Fig. 2.2-2, using which liquid and gaseous insulating materials can be made to break down at a gap spacing of 2.5 mm.

---

[1]) Comprehensive survey, e.g. in [*Holzmüller, Altenburg* 1961; *Imhof* 1957].

[2]) Detailed information in [DIN 53481 and VDE 0303 Part 2/11.74].

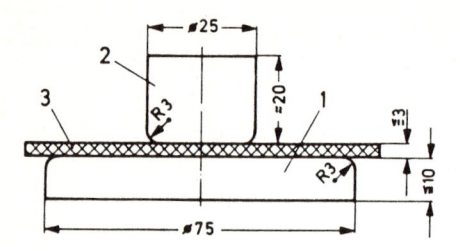

Fig. 2.2-1 Plate-plate electrode arrangement for measurement of the breakdown voltage of solid insulating materials for materials up to 3 mm thickness.
1,2 plate electrodes
3 insulating material sample

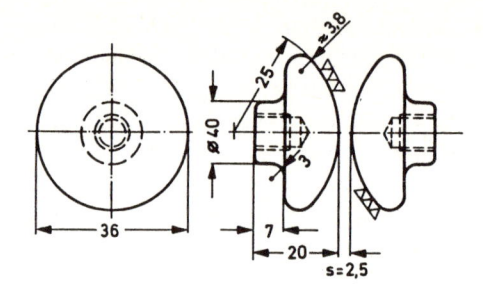

Fig. 2.2-2 Electrode arrangement with spherical segments for measurement of the breakdown voltage of liquid insulating materials

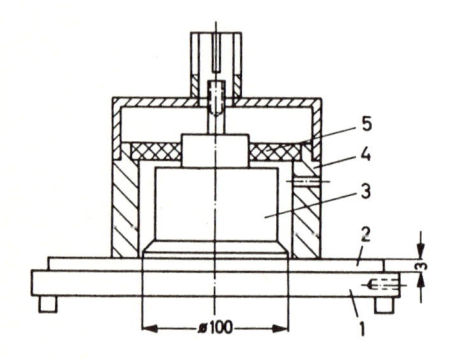

Fig. 2.2-3
Plate electrodes with guard ring for measurement of the volume resistance of solid insulating materials
1 live plate
2 insulating material sample
3 measuring electrode
4 guard ring
5 insulating and guiding part

The breakdown test is conducted with alternating voltage, which should be increased from zero to the breakdown value within 10...20 s. The median value of the breakdown voltage is determined from 5 samples; if any value lies more than 15% off the median value, 5 additional samples must be tested and the median value then determined from 10 samples. The breakdown field strength can be evaluated from the breakdown voltage and the smallest electrode spacing.

b) Insulation resistance[1])

Practical insulation systems frequently comprise many dielectrics which are stressed in parallel. Thus, for example, the insulation resistance of a support insulator consists of the combination in parallel of surface resistance and volume resistance. While the volume resistance, commonly expressed as specific resistance in $\Omega\,cm$, is often independent of the surrounding medium, the surface resistance is appreciably influenced by ambient conditions such as pressure, temperature, humidity, dust etc.

An arrangement for the measurement of the volume resistance of plate-type insulating material samples is shown in Fig. 2.2-3. The live electrode, which also supports the plate-type sample, is arranged opposite a measuring electrode. The volume resistance is calculat-

---

[1]) For detailed information about testing arrangements, see [IEC Publication 93 (1980) and 167 (1964)].

ed from the direct voltage applied (100 V or 1000 V) and the current taken by the measuring electrode. A guard ring arranged concentrically around the measuring electrode with a 1 mm gap prevents erroneous measurements caused by surface currents.

Special testing arrangements are available for tube-shaped insulating material samples, for insulating compounds which can be melted and for liquid insulating materials.

Common insulating materials exhibit a specific volume resistance of $10^{12}...10^{13}$ Ωcm, whereas superior materials can reach resistance values up to 10^{17} Ωcm or even higher.

To measure the surface resistance metallic knife-edges are used, set up at a gap spacing of 1 cm over a length of 10 cm on the surface of the insulating material under test and direct voltage is applied. From the voltage and current, the surface resistance, expressed in ohms, is determined.

c) Tracking strength [1])

When an insulation system is electrically stressed a current which is determined by the surface resistance flows on its surface and is referred to as leakage or creepage current. It is easy to understand that the ambient conditions temperature, pressure, humidity and pollution essentially determine the magnitude of this leakage current. Technical insulating materials must be resistant to this leakage current, i.e. no, or only limited, deterioration of the surface properties shall occur.

Leakage currents result in thermal, and due to the by-products, also chemical stressing of the surface. The visible effects of overstressing are tracks resulting from material decomposition; these can appear in the form of a conducting path making further electric stressing of the material impossible, or as erosion, without leaving a conducting track behind. Although the insulating properties are adversely affected by erosion, e.g. by the ease of dust deposition, yet further electric stressability is not precluded. Erosion occurs either in plates or as pits (Fig. 2.2-4).

Tracking is not restricted to insulating surfaces of outdoor arrangements but can also occur, under unfavourable circumstances, in indoor applications or even in the interior of equipment. It is influenced for instance by the material properties, by the form and finish of the electrodes and the surfaces, and also by the external conditions. Tracking is caused by a mechanism similar to the one described in Section 1.6 for pollution flashover. Due to merging of several localized tracks, a complete flashover can be facilitated or initiated.

Testing of insulating materials for their tracking strength is undertaken according to prescribed methods. In the method KA according to VDE platinum electrodes are placed on the insulating material sample of at least 3 mm thickness and 380 V alternating voltage is applied to the electrode arrangement as shown in Fig. 2.2-5. A pipette provides one drop of the testing solution of prescribed conductivity every 30 s, this wets the surface of the insulating material between the electrodes and causes leakage currents. As test result either the number of drops up to the time of automatic switch-off of the test circuit is evaluated, or the greatest pitting depth is measured.

---

[1]) For detailed information, see [IEC Publication 112 (1979)].

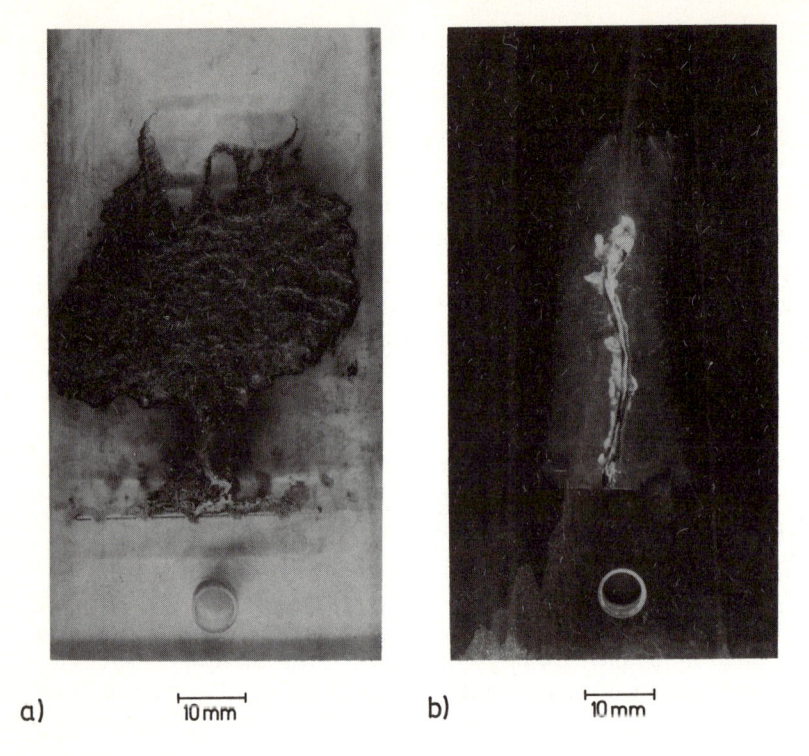

Fig. 2.2-4 Plate a) and pitting b) type of erosion on epoxy resin moulds

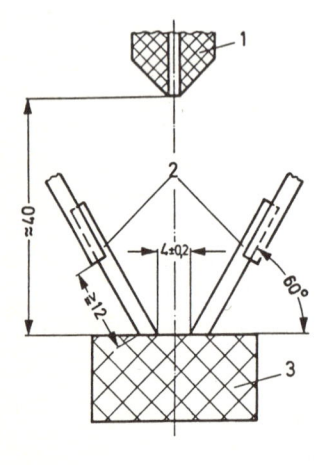

Fig. 2.2-5
Arrangement for determination of tracking strength
1 pipette
2 platinum electrodes
3 insulating material sample

2.2 Properties and testing of insulating materials

d) Arcing resistance[1])

Flashovers along the surfaces of insulating materials, with a subsequent power-arc, are indeed very rare, but basically unavoidable faults in technical insulation systems. Insulating materials exposed to the influence of the arc must therefore experience no, or only minimal, variations in their electrical and mechanical properties, viz. be arc resistant. Due to the high arc temperature and as a consequence of incomplete burning of the insulating material, conducting tracks can remain, which no longer permit further electric stressing.

To determine the arcing resistance, carbon electrodes supplied with 220 V direct voltage are set up on the insulating plate. With an arc struck on the surface of the insulating material, the electrodes are moved apart at a velocity of 1 mm/s up to a maximum separation of 20 mm. Six levels of arcing resistance, L1 to L6, defined according to the destruction caused by the arc, are used to judge the materials.

e) Dielectric constant and dissipation factor[2])

A dielectric constant ϵ_r different from 1 is caused by polarization effects in the insulating material. For practical insulating materials, apart from the deformation polarization (electronic, ionic, lattice polarization), the orientation polarization is of particular significance since many insulating materials have permanent dipoles in their molecular structure. This is the major cause of polarization losses and is responsible for the frequency dependence of ϵ_r and $\tan \delta$, which is important for technical applications.

Since the various polarization mechanisms possess different relaxation times, the variation of ϵ_r as a function of frequency is as shown schematically in Fig. 2.2-6. The different relaxation times result in frequency limits beyond which the respective mechanisms no longer exist, because the corresponding dipole movement does not occur. This is why the dielectric constant must also decrease. With a change of state, step-like variations of ϵ_r can occur on account of the changed mobility of the dipoles.

At each transition region of the dielectric constant the dissipation factor $\tan \delta$ has a maximum. But only the transition region from a to b (Fig. 2.2-6) is interesting for technical insulation systems, viz. the frequency range in which orientation polarization vanishes.

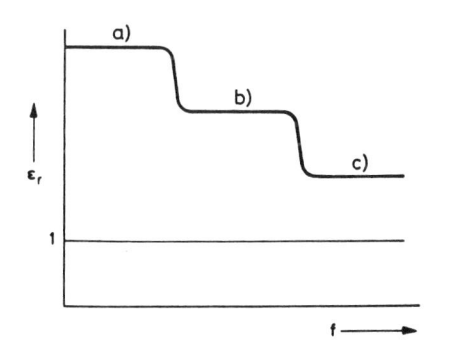

Fig. 2.2-6
Schematic representation of the frequency dependence of the dielectric constant
a) orientation polarization
b) ionic polarization
c) electronic polarization

---

[1]) For details see [DIN 53484 and VDE 0303 Part 5/10.55].
[2]) Comprehensive treatment in [Holzmüller, Altenburg 1961; Anderson 1964; Hippel 1958].

Important statements about the properties of an insulating material result from dependence on voltage and temperature [*Kind* 1972]. If the curve $\tan\delta = f(U)$ shows an ionization knee-point, it proves the onset of partial discharges. The occurrence of polarization losses, as well as the increase of losses due to ionic conduction are recognized from the curve $\tan\delta = f(\vartheta)$.

Measurement of $\tan\delta$ and the determination of ϵ_r are done, as is well-known, using bridge circuits.

2.2.2 Thermal properties

In equipment and installations for the supply of electricity, heat is generated by ohmic losses in conductors, through dielectric losses in insulating materials and through magnetization and eddy-current losses in the iron. Since, by comparison with metals, insulating materials have only a very low thermal stability, the permissible temperature rise of the insulating material often restricts the use of the equipment. Knowledge of the thermal properties of insulating materials is, therefore, important for the construction and design of equipment and setups.

a) Specific heat

Due to the inertia of thermal transport, an insulating material must be in a position to absorb short-duration thermal pulses, caused by rapid load variations, via its thermal capacitance by an increase in temperature. The specific heat c of a few important materials is compiled in Table A 3.1 of Appendix 3. For adiabatic heating we have:

$$\Delta T = \frac{W}{c \cdot m},$$

where m is the mass and W is the supplied energy.

b) Heat transport

During continuous stressing under static operating conditions, the heat generated as a result of loss must be dissipated through the surroundings. Transport mechanisms are thermal conduction, convection and radiation [*van Legen* 1971]. In thermal conduction the thermal current flowing between flat plates is expressed by

$$P = \frac{A}{s} \cdot \lambda (T_1 - T_2),$$

where A is the area of the plate, s is the plate thickness and $(T_1 - T_2)$ is the temperature difference.

The proportionality factor λ is the thermal conductivity which can be assumed to be constant in the technically interesting temperature range; it is listed for a few important materials in Table A 3.1 of Appendix 3.

For rapid removal of loss heat from electrical equipment good thermal conductivity is desirable. This requirement is satisfied best by crystalline insulating materials because the regular arrangement of atoms in the crystal lattice and the small atomic spacings ensure good transmission of atomic movement. In contrast, amorphous materials have a distinctly poorer thermal conduction, as is clear in the example of crystalline and amorphous quartz. While for quartz crystal $\lambda = 6...12$ W/mK, for quartz glass it is $\lambda = 1.2$ W/mK.

2.2 Properties and testing of insulating materials

The good thermal conduction properties of quartz effect a noticeable increase of λ in the case of filled mouldings, when crystalline quartz in the form of sand or quartz powder is used as filler material.

For heat transport by convection the thermal current P is proportional to the boundary area A and the temperature difference between the dissipating and absorbing medium:

$$P = \alpha \cdot A(T_1 - T_2).$$

The thermal transition number α is not a material constant but depends upon parameters such as density and specific heat of the flowing medium, the velocity of flow and the type of flow. For preliminary calculations the following values can be adopted [*Oburger* 1957]:

	α in W/m² K
fixed object/stationary air	3.4 ... 35
fixed object/moving air	12 ... 600
fixed object/liquid	250 ... 6000

Since the values each cover a wide range, in practical cases it is necessary to calculate exact values of α using the literature [VDI 1974].

Thermal transport by radiation, not treated here in detail, is only significant in the case of circuit breakers and SF_6-installations.

c) Linear thermal expansion

Insulating materials are construction materials which are frequently employed in contact with metals. On account of the larger thermal expansion of organic insulating materials, the danger of mechanical overstressing exists and this could result in the development of cracks or electrode detachment. For inorganic insulating materials the linear thermal expansion is lower than for metals; so an improvement is effected by filling organic materials with inorganic substances e.g. epoxy resin with quartz. Partly crystalline materials very often have a greater thermal expansion than amorphous materials (Table A 3.1 in Appendix 3).

d) Thermal stability

An important property of insulating materials is the shape retention on heating[1]); there are two methods to determine this. The warm shape retention according to *Martens* is determined for a standard testing rod of 10×15 mm² cross-section and 120 mm length, which is stressed uniformly over the entire length with a bending stress of 500 N/cm². At the same time the temperature of the surrounding air is increased at the rate of 50 °C/h. The temperature at which the rod reaches a specific bending denotes the warm shape retention according to *Martens*. For thermoplastic materials the *Vicat* method is also applied. The *Vicat* temperature is that temperature at which a blunt needle of 1 mm² cross-section under a force of 10 N or 50 N penetrates to a depth of 1 ± 0.1 mm into the insulating material. The following table contains some relevant data:

---

[1]) For details see [DIN 53458].

Material	shape retention according to *Martens* in °C	under heat according to *Vicat* in °C
PVC	60	70... 90
PTFE	70	75...100
EP-moulding	up to 160	–
PUR	up to 80	–
PE	–	40... 75

In the plastic range, the mouldings not only suffer a marked decrease in tensile, compressive and bending strengths but also a noticeable deterioration of their electrical and dielectric properties.

A high value for shape retention under heat is a decisive advantage of inorganic insulating materials over the organic kind.

2.2.3 Chemical properties

When foreign matter diffuses into insulating materials, this can cause chemical change. Only inorganic materials such as glass and densely fired ceramic materials are practically impermeable. In synthetic organic materials diffusion can take place within the molecular framework of the polymer [*Brinkmann* 1975]. The diffusion velocity depends upon the material structure and the affinity of the base material for the foreign substance.

For example, all organic insulating materials absorb moisture by diffusion. The dielectric and electrical properties deteriorate as a result. Dissolved salts produced by hydrolysis or from impurities increase the conductivity and cause a poorer dissipation factor and breakdown field strength. The high dielectric constant of water modifies the dielectric constant of the material and causes a change in the voltage distribution during alternating voltage stressing. In addition, the absorbed water can lead to changes in dimension (bloating) and corrosion of the electrodes.

Insulating materials for outdoor application should have poorly wettable surfaces so that closed water paths, which reduce the strength, are avoided. The wettability is characterised by the angle of contact on technically clean surfaces shown in Fig. 2.2-7. The larger the angle ϑ_{max} in the direction of motion becomes, the lower the wettability. The following table shows extreme values:

Insulating material	ϑ_{max}	ϑ_{min}
paraffin	110°	95°
silicon-rubber	100°	90°
glass, mica	0°	0°

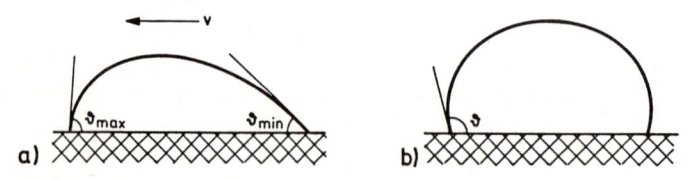

Fig. 2.2-7 Angle of contact of insulating materials
a) moving drop   b) drop at rest with angle of contact > 90° (e.g. water on PTFE)
v   direction of motion

Inorganic materials e.g. porcelain and glass, show resistance to alkalis and acids (with the exception of hydrofluoric acid); organic materials are attacked by strongly oxidizing acids, by alkalis as well as by hydrocarbons specific to the material. In outdoor application of insulating materials wet pollution layers can be decomposed by the electric stress and the heat so generated to form agressive chemicals; these, with the additional influence of light, oxygen, ozone, heat and UV-radiation, attack the insulating material.

2.3 Natural inorganic insulating materials

Natural, that is not chemically prepared, inorganic materials are used as high-voltage insulating materials in the form of natural gases and a few solid materials of mineral origin, such as mica and quartz.

2.3.1 Natural gases

Important gases in high-voltage technology applications are air, nitrogen, hydrogen and helium, whereas natural gases such as oxygen, carbon dioxide, neon, argon, krypton and xenon are of minor significance.

Nitrogen and air possess the highest breakdown field strengths of all natural gases. Under standard conditions in a homogeneous field, a breakdown field strength of about 30 kV/cm can be expected. As shown in Appendix 2, the electric strength also depends upon the electrode geometry.

In order to increase the breakdown field strength compressed gas is employed. When using air the danger of oxidation always exists, so dry nitrogen is often used, as for example in compressed gas capacitors, Van-de-Graaff generators operating in pressure tanks, internal and external gas pressure cables. In the high-pressure region, the electric strength increases nearly proportionally with the pressure. However, at pressures in excess of about 10 bar, irregularities of the electrode surfaces increasingly determine the breakdown behaviour.

Liquid nitrogen (boiling point 77 K) is suitable as an insulating or impregnating medium for low-temperature insulation systems (e.g. [*Peier* 1976]).

Hydrogen is not used as an insulating gas, but, due to its good thermal conductivity, it is important in the cooling of large electrical machines. In low oil content circuit breakers the switching arc disintegrates the oil whereby hydrogen in particular is released which, by dissociation at ca. 4000 K, absorbs a large amount of heat and so cools the arc. Whereas the thermal conductivity of hydrogen at room temperature is 0.18 W/mK, it increases to about 50 W/mK at 4000 K and so is many times greater than that of other quenching media.

Whilst the inert gases have acquired some significance for filling incandescent lamps and strip-lighting, liquid helium (boiling point 4.2 K) is to this day the irreplaceable cooling medium for superconducting installations. On account of its remarkably good electric strength in the temperature range below 100 K, it can also be used as electrical insulation in superconducting equipment [*Gerhold* 1977].

A few properties of natural gases are summarized in Table A 3.2 of Appendix 3, where, for purposes of comparison, sulphurhexafluoride (SF_6) has also been included.

2.3.2 Quartz and mica

In its pure form quartz (SiO_2) is found in rock crystal and as quartzite in sand; in addition to the crystalline form, it also occurs in amorphous form, e.g. in quartz glass. Its dominant property is the high volume resistance at high temperatures; at 500 °C, for example, it is $5 \cdot 10^9$ Ω cm, and at 1000 °C still as much as 10^6 Ω cm. Quartz is therefore commonly used in high-temperature insulation systems, as in electrostatic precipitators, in which the insulation must be capable of withstanding flue gas temperatures of up to 400 °C. Quartz sand is used in fuses for arc quenching purposes and quartz powder is used as a filler material for epoxy resins.

Mica is a natural mineral, known under the name muscovite if it contains potassium or phlogopite if it contains magnesium. The crystalline structure is a sheet structure with very strong bonds in one plane and very weak Van-der-Waal's forces in the plane normal to this, so that mica can be split easily into fine flakes; the flake thickness is of the order of 0.02...0.1 mm. For high-voltage insulations the mica flakes are transformed into mechanically stable plates or pipes (micanite) by the use of bonding materials e.g. shellac or EP-mouldings; or glued, the flakes overlapping one another, on substrate bands of paper or fibre-glass with bonding materials, forming flexible tapes (micafolium) which can be further processed. Since the mica flakes overlap, air-pockets, and hence partial discharges, are unavoidable.

The important properties of mica are excellent tracking strength; high breakdown field strength; high partial discharge stability and high volume resistance; a dissipation factor independent of frequency up to 50 MHz and, in particular, long-term thermal stability up to 600 °C. Mica products generally possess poorer properties than pure mica due to the influence of the bonding materials used. Micanite with low bonding material content is used for commutator insulation in electric machines and micafoil with epoxy resin impregnation for the insulation of machine windings.

2.4 Synthetic inorganic insulating materials

Materials in this group of interest to high-voltage insulation systems are sulphurhexafluoride, glass, ceramics and ceramic oxides. Whereas amongst the ceramic materials porcelain and steatite are significant, amongst the metal oxides aluminium oxide in particular should be mentioned.

2.4.1 Sulphurhexafluoride (SF_6)[1]

Among the manifold synthetic gases, and particularly the gaseous halogen compounds, SF_6 is, to this day, unchallenged as an insulating gas. It possesses high electric strength which, under otherwise similar conditions, is about 2.5 times that of air, as well as excellent arc quenching properties.

Sulphurhexafluoride is obtained from molten sulphur and gaseous fluorine at 300 °C and finally refined to 99.9% purity. It is a colourless, odourless and non-poisonous gas, chemically inert and non-inflammable. Under the influence of high temperatures (arc), however, poisonous by-products can be formed. In moist gas hydrofluoric acid can form,

---

[1]) Summary in [*Mosch, Hauschild* 1978].

2.4 Synthetic inorganic insulating materials

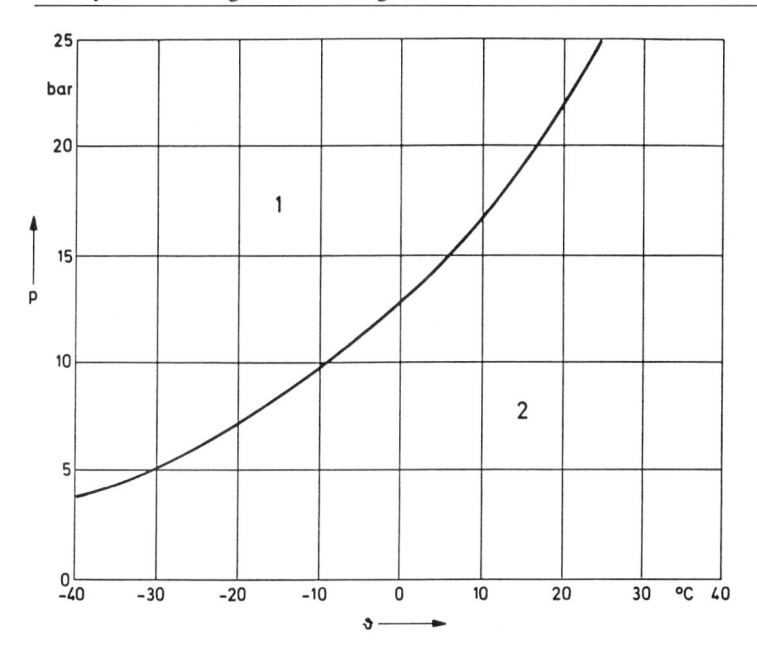

Fig. 2.4-1 Vapour pressure curve of SF_6 1 liquid state   2 gaseous state

so that the use of insulating materials such as glass and porcelain which are sensitive to hydrofluoric acid should be avoided in SF_6. It is, as Fig. 2.4-1 shows, easily liquified so that the working pressure of SF_6 insulated equipment exposed to the risk of low outdoor temperatures should not be chosen too high, or provision must be made for heating.

SF_6 has good thermal stability and decomposes at temperatures above 800 K only. At temperatures above 2000 K it dissociates completely. The low thermal conductivity of the heavy fluorine and sulphur atoms causes a high arc temperature which, in conjunction with the very low ionisation energy of elementary sulphur, leads to high electrical conductivity of the arc. Low arc voltage and small energy dissipation in the arc are the results. At the boundary region of the arc S and F atoms recombine to form SF_6, releasing the dissociation energy absorbed from the arc core to the cooler environment. The favourable radial temperature distribution of an arc in SF_6 results in the dissociation temperature again being undercut in the arc core, only a few μs after current zero; the entire gas volume recovers its electron affinity, resulting in rapid dielectric recovery of the gap [*Erk, Schmelzle* 1974]. As in all gases, the electric strength also depends on the geometry of the configuration. Breakdown field strengths of SF_6 for plate, cylindrical and spherical electrodes are compared with those of other gases in Appendix 2.

The outstanding arc quenching properties are utilized to advantage in SF_6 circuit breakers. The high breakdown field strength has made the development of metal-clad SF_6 insulated substations possible, which, compared with outdoor substations, require only a fraction of the latter's space; they are therefore installed to particular advantage in densely populated areas, and also in regions with an increased pollution risk. The pressure range of these installations lies between 1.5 bar and 5 bar. Finally, the SF_6-insulated cable promises convenient technical solutions for transmission of high power in the GW-range, over short and medium distances.

2.4.2 Glass[1])

Glass is the oldest known insulating material which was already made use of by *R. Boyle* in his vacuum amber experiments in 1675. The most well-known historical application, however, are the Leyden jars.

As a supercooled liquid of high viscosity, glass is an amorphous material; on solidifying crystallization does not occur, but the irregular structure of the liquid is retained. It is manufactured by melting various oxides, when, depending upon its application, different mixing proportions are set [*Oburger* 1957]:

- SiO_2 silicon dioxide in the form of quartz sand, as main component up to 70%
- B_2O_3 boron trioxide, up to 16% improves the electrical properties and temperature change durability, reduces the expansion coefficient
- Al_2O_3 alumina 0.5...7%, improves weather durability, reduces the expansion coefficient
- PbO   lead oxide, increases the electrical resistance
- BaO   barium oxide
- CaO   calcium oxide.

Alkalis, e.g. soda (Na_2CO_3), in amounts of 2...18%, help reduce the melting temperature, which for pure quartz lies at 1700 °C. Glass for electrotechnical applications (E-glass) must be poor in alkali content in order to achieve low conductivity, i.e. should be assigned an alkali content of less than 0.8%.

The properties of glass are summarised in Table A 3.3 of Appendix 3. Good tracking strength and breakdown field strength compare with a tolerable dissipation factor and volume resistance, but the latter decreases rapidly with increasing temperature (ionic conduction). Water absorption by glass in nil, yet positive sodium ions on the glass surface are easily leached. Caution is recommended during direct voltage stressing because of glass electrolysis, which, by migration of positive ions to the cathode can lead to alkali depleted layers with changed physical properties. The high long-term thermal stability is of advantage.

The processing of standard glass is by fusing (e.g. cable terminal housing of borosilicate glass in the medium voltage range) or by compression (e.g. in cap-and-pin type glass insulators). During manufacture of cap insulators the molten glass at a temperature of 1050 °C is fed to rotating presses and pressed into discs. Internal tension arising from asymmetrical cooling is removed by tempering, after which controlled cooling is undertaken using an air-blast. In this way the outer skin of the glass disc, to a thickness of a few mm, acquires a pre-stress which determines the mechanical strength of the finished insulator. However, in so doing the glass cap also becomes sensitive to internal faults and to minor damage of the pre-stressed layer; extensive temperature shock tests and longer storage time are therefore necessary before installation.

---

[1]) Survey in [*Brinkmann* 1975].

2.4 Synthetic inorganic insulating materials

The particular application of E-glass is as glass fibres for fibre reinforced plastic materials. Manufacture takes place by the jet draw method (Fig. 2.4-2). The glass melt 1 is pressed through the perforated nipple 2; glass fibres of about 10 μm diameter pass through a planishing bath with velocities up to 60 m/s and are then wound as a filament with 100...1000 individual threads. The planishing bath provides a surface coating for the sensitive, freshly drawn threads; for electrotechnical applications a planishing bath should be used which guarantees flawless, i.e. electrically and mechanically strong, attachment to the resin matrix.

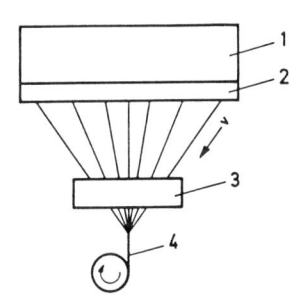

Fig. 2.4-2 Jet draw method for manufacture of glass fibres
1 glass melt   2 perforated nipple
3 planishing bath 4 filament
$v \simeq 60$ m/s

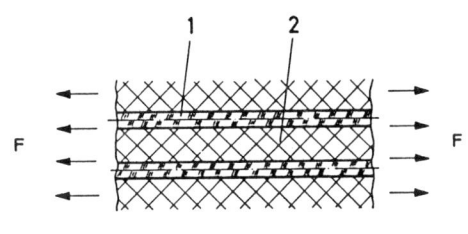

Fig. 2.4-3 Glass fibre in a resin matrix
1 glass fibre   2 resin matrix
F tensile force

The properties of E-glass fibres are also listed in Table A 3.3 of Appendix 3. The extraordinarily high tensile strength is attributed to the small diameter and the surface tension; this is the basic requirement for good mechanical properties of glass-fibre reinforced plastics.

Glass fibres are either processed to filament with axially parallel arrangement of individual fibres, to glass-fibre mats with an irregular lamination of ca. 50 mm fibre length, or to glass-silk fabric.

Glass fibres are used wherever high tensile strength is demanded in conjunction with good electrical insulation, e.g. for

    tying and bandaging of windings in electric machines,

    bandaging core packets of transformers,

    resin impregnated fibre-glass thread as supporting core for compound insulators,

    resin impregnated fibre mats or pieces of fabric as insulating plates.

The good mechanical properties of fibre-reinforced plastics are explained by the almost exclusive assumption of the forces by the glass fibres. Let a glass fibre under tension be embedded in impregnating resin as shown in Fig. 2.4-3. For the mechanical tensions in glass and resin, σ_G and σ_H, under tensile strain we have:

$$\frac{\sigma_G}{\sigma_H} = \frac{E_G}{E_H}$$

where E_G, E_H are the elasticity moduli of glass and resin respectively.

The characteristic data of fibre and e.g. EP-moulding are:

	E modulus kN/mm²	Tensile strength N/mm²
Glass fibre	70	2500
EP-moulding	2.5...4	60...80

Hence the tension σ_G is 15 to 30 times the tension σ_H and the glass fibre assumes the mechanical load practically alone, which is desirable regarding its high strength. Since transmission of the force from one fibre to the next takes place with the participation of the resin matrix, particular significance is assigned to the strong boundary surface bonds. This is also important during electric stressing in the direction of the fibre (e.g. in long rod compound insulators) where insufficiently strong bonding between fibre and resin could cause electrical tracking phenomena and initiate breakdown along the fibre.

2.4.3 Ceramic insulating materials

Ceramic insulating materials are products of inorganic raw materials, especially silicates and oxides, which are formed as a raw material mixture and obtain their properties by sintering. They comprise crystal structures embedded in a glass matrix.

A raw material mixture of defined fineness (grain size distribution) is obtained from natural or artificial inorganic raw materials; this mixture is processed further either in a dry or wet state during shaping. After drying the irreversible formation of the ceramic material is achieved by firing (sintering). Since drying and sintering cause shrinkage a final finishing process is often necessary.

The advantage of ceramic materials lies in the easy processability and workability of the raw material mixture before firing, for which reason there appears to be an unlimited abundance of shapes for ceramic products; however, in cases where final finishing is necessary, the high degree of hardness of the material is a disadvantage.

Appropriate to their importance in high-voltage insulation technology, the ceramic materials porcelain, steatite and aluminium oxide (alumina) will now be subjected to closer examination.

a) Porcelain and steatite

Porcelain is an aluminium silicate. Processing the substance in the unfired state requires that it have a certain plasticity, which is given by the plastic raw materials; in addition, there are the non-plastic or hard materials.

A classical raw material composition is:

   50% kaolin, 25% felspar, 25% quartz.

Due to the high proportion of quartz, the resultant material is designated quartz porcelain. The values of characteristic properties correspond to KER 110.1 as in DIN 40685.

Kaolin, which to some extent is replaced by clay, contains clay minerals in the form of aluminium hydrosilicates, e.g. kaolinite ($Al_2O_3 \cdot 2SiO_2 \cdot 2H_2O$). On firing, kaolinite decomposes into quartz (SiO_2) and corundum (Al_2O_3) in the crystalline state and releases

2.4 Synthetic inorganic insulating materials

the water of crystallization. Felspar forms the glass phase after sintering, that is the glassy base substance of porcelain, with about 50% volume proportion. Felspar is often introduced as potash-felspar.

At the high temperatures in the glass phase quartz partly goes into solution but crystallises out again on cooling. Together with the Al_2O_3 crystals, the crystal group and size of the quartz crystals and their embedding in the glass matrix exert a decisive influence on the material properties.

Nowadays predominantly alumina porcelain according to KER 110.2 as in DIN 40685 is used. Here quartz is partly or wholly replaced by Al_2O_3 which yields greater mechanical strength as well as better workability during manufacture. A typical composition is e.g.

    40...50% plastic components (kaolin and clay)
    30% felspar
    30...20% alumina.

Due to the Al_2O_3 content of the plastic raw materials, the total content of Al_2O_3 in clay porcelain of this kind lies at 50%. Alumina is extracted from bauxite and introduced as a processed raw material.

Steatite is a magnesium silicate. The original raw material is soapstone, a magnesium hydrosilicate $(Mg_3Si_4O_{10}(OH)_2)$, and at 85% represents the major component; in addition 10% clay and 5% felspar are also included. The values of characteristic properties correspond to KER 220 as in DIN 40685.

b) Manufacture of porcelain, based on the example of high-voltage insulators

The hard raw materials including the clays are powdered in crushers and ground wet in ball-mills until a particular grain size distribution is achieved. The introduction of kaolin then takes place in stirring vessels so that a fluid raw material mixture with about 50% water content, the so-called slip, results. After partial removal of water in mechanical filter presses and intermediate storage, initial shaping into cylindrical bodies follows by vacuum extruders. Vacuum pre-treatment of the plastic material is essential to ensure that gas occlusions are prevented; these would cause a drastic reduction in the strength of the fired sample. After intermediate drying to about 12% water content the cylindrical bodies are brought into the final raw shape by turning. Subsequent drying removes the physically bound water down to about 0.5% content, the procedure causing a shrinkage of about 8%. The dried sample is then coated with a ceramic suspension, the glaze, which on firing produces a glassy-smooth surface of the desired colour.

During subsequent firing, which is undertaken in continuously operating tunnel kilns or discontinuously driven chamber kilns or batch-cart kilns, the ceramic material is formed by sintering. On account of the loss of water of crystallization and the occlusion of open pores, a so-called firing-shrinkage, also of about 8%, occurs. So the overall shrinkage due to drying and sintering lies between 15 and 20%; this must be taken into account by dimensional tolerances when shaping in the raw state.

Firing takes place at about 1400 °C. Since the material then has only poor mechanical strength and shape retention, special measures are necessary to ensure shape stability (firing in a hanging or standing position, guidance).

In the kilns, which are sometimes several meters high and equally wide, the temperature distribution over the cross-section is not constant. The material must therefore possess a certain temperature range — the sintering range — in which compact sintering occurs.

If the temperature remains below the minimum temperature, sintering is not complete and the material remains porous; if the temperature exceeds the maximum value, overfired material with secondary porosity results. The advantage of clay porcelain in comparison to quartz porcelain, and especially steatite, lies in its large sintering range and its comparatively high strength at the firing temperature.

Insulating structures which require exact working dimensions need processing after firing, using the saw, grindstone or drill. Here, due to the hardness of the material, finishing is usually only possible using diamond tipped tools.

Before assembling the metallic terminal fittings by lead sealing or cementing with Portland cement, mechanical tests as well as an ultrasonic test are usual. With the aid of the latter internal defects (shrinkage cavities, cracks) which would reduce the mechanical strength can be recognised non-destructively. Many tests must be repeated after assembly is complete or, like the tensile test on long-rod insulators, can only be performed then at all.

c) Properties and application of porcelain and steatite

A few important properties of quartz porcelain, clay porcelain and steatite are listed in Table A 3.3 of Appendix 3. The raw density quoted takes into account that the materials named, even after compact sintering, possess a pore volume of 2...6%. However, this pertains to occluded and hence harmless pores; exposed porosity, on the other hand, permits the infiltration of water and causes mechanical failure as a result of material fatigue after the action of frequent frosts.

Porcelain and steatite are gas-tight, light and corrosion proof, chemically inert to all alkalis and acids with the exception of hydrofluoric acid, and therefore particularly resistant to contamination in outdoor applications, thermally stable and arc-resistant. A certain sensitivity to local mechanical over-stressing does exist however, on account of the brittleness of the material; fracture occurs spontaneously without previous flow.

The advantages of steatite are the consequence of the high mechanical strength and the low dissipation factor. Steatite is, therefore, specially suited to high-voltage high-frequency applications as in tower bases and shackle insulators for antennae. Porcelain, particularly in the form of clay porcelain, dominates the outdoor insulation of high-voltage equipment. Transmission line insulators, traction insulators, pressure resistant switchgear porcelain, post insulators for isolating switches and busbars are all made of this material as are transformer bushings, housing for current and voltage transformers, coupling capacitors and lightning arrestors.

d) Alumina

Ceramic oxides have a wide field of application, and in high-voltage technology alumina is the most important. Thus in vacuum breakers and thyristor enclosures its excellent electrical and mechanical properties are exploited and also its aptitude for application of hard solder metallizations. Moreover, the good thermal conductivity and good insulating properties at high temperatures are noteworthy: the specific resistance, even at 1000 °C, is still 10^6 Ω cm.

2.5 Natural organic insulating materials

Organic materials are characterized by the element carbon which has the ability to form long chains or ring-type structures. Mineral and vegetable oils belong to the group of natural organic liquid materials; paraffin and bitumen as crude-oil products belong to the solid group as do in addition, wax, resins (e.g. shellac), wood and fibrous materials like paper, silk, cotton and jute. Important to high-voltage technology are mineral oil, paper, and, with some limitations, wood and bitumen.

2.5.1 Mineral oil

Mineral oil is obtained by fractional distillation of crude oil after degassing, dehydrating and desalination of the raw product. Mainly saturated hydrocarbons with naphthene or alkane structure are employed since they are chemically more stable than the unsaturated aromatic hydrocarbons. A few examples of the structures are given below:

pentane (alkane): $CH_3-CH_2-CH_2-CH_2-CH_3$

cyclohexane (naphthene):   benzene (aromatic):

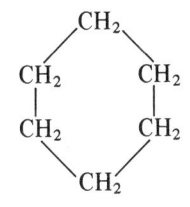

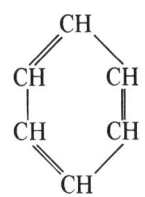

Removal of the undesirable aromatic hydrocarbons is achieved by a process, subsequent to the distillation, referred to as refining.

As shown in Section 1.4 the electrical properties of insulating oil deteriorate with increasing water and gas content. Insulating oil must therefore be pre-treated before its use in high-voltage equipment. This is done in refining plants for degassing and desiccation [e.g. *Beyer* 1971]. Surface degassing is usually adopted where a thinly flowing film of large area is produced. During this procedure the oil is exposed to a temperature of 50 to 60°C in a vacuum of about 10^{-2} mbar. Fig. 2.5-1 shows the basic setup of an oil refining plant. In the new state viz. refined, insulating oil should have a breakdown voltage of 50...60 kV with the electrodes shown in Fig. 2.2-2, this corresponds to a breakdown field strength of about 200 kV/cm.

Fig. 2.5-1
Principle setup of an oil-refining plant
1 oil storage tank
2 circulating pump
3 degassing chamber
4 vacuum pump
5 basket with Raschig rings

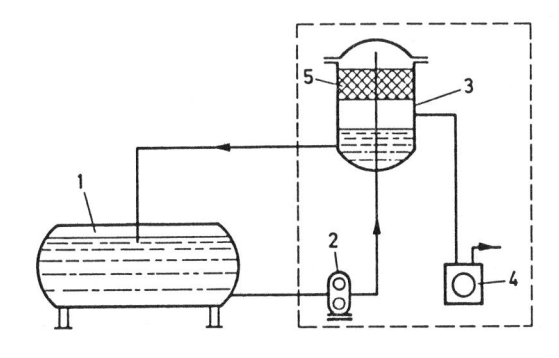

Insulating oil is subject to ageing risk due to absorption of moisture, the solution of gas, impurities and, in particular, oxidation. Under the combined effect of oxygen and heat, oxidation products are formed which are soluble in oil, e.g. acids, and insoluble components which appear as sludge. The oxidation of oil is accelerated by the catalytic action of copper, which is the reason why bare copper conductors must be avoided in insulating oil. The neutralization number and the saponification number are useful to characterize the ageing state, the former gives the quantity of potassim hydroxide (KOH) necessary to neutralize the free acids contained in 1 g of oil, while the latter describes the quantity of KOH which neutralizes the free and bound acids, and so incorporates the neutralization number as well.

As Fig. 2.5-2 shows, the dissipation factor of an aged oil is about ten times worse than that of new oil [*Oburger* 1957]. Large oil-insulated apparatus, e.g. transformers, must therefore be regularly checked for the ageing condition of the insulating oil. For this purpose oil samples are withdrawn and examined for their breakdown field strength, dissipation factor and impurities. If necessary, refinement must be undertaken on site or total replacement of the oil carried out. Replacement of the oil is recommended when the neutralization number exceeds 0.5 mg KOH/g oil, or when sludge soluble in chloroform is observed [*Müller, Molitiv* 1977].

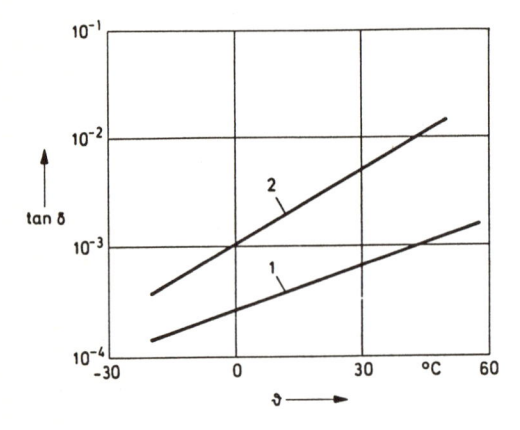

Fig. 2.5-2
Effect of ageing on the dissipation factor of transformer oil
1 new oil 2 aged oil

In addition to the method described in Section 3.1.4 for the prevention or reduction of ageing of oil, "inhibition" of mineral oil is also practised. Here, the resistance to oxidation is improved by the addition of ageing inhibitors. These react with the broken bonds in the oil molecules and interrupt the oxidation process by forming stable, inert and dielectrically harmless compounds. They are used up in the process and must therefore be replenished from time to time [*Brinkmann* 1975].

Some properties of low viscosity mineral oil in refined condition, such as it is used as transformer oil in devices or for oil-filled cables, are listed in Table A 3.4 of Appendix 3. The transparent, clear liquid has a solidifying point at $-40\,°C$. The electrical properties depend upon the purity. The quoted breakdown field strength of 25 kV/mm is valid for spacings in the mm-range; for thin layers breakdown field strengths up to 100 kV/mm have been measured and for films in the μm-range a value of 300 kV/mm has been quoted

2.5 Natural organic insulating materials

[*Brinkmann* 1975]. Combined with the low dielectric constant, this is the reason for the excellent electric strength of oil-paper insulation systems (see Section 2.5.3).

Under the action of electric discharges oil decomposes with the formation of gas. In arcs thermal degradation of the oil occurs with the formation of about 60% hydrogen, ca. 10% other gases and ca. 25% saturated and unsaturated hydrocarbons. The high hydrogen content of the degradation products results in concentrated cooling of the arc [*Brinkmann* 1975].

At sharp-edged electrodes, e.g. at the edges of metal foils in capacitors, continuous discharges can also cause polymerization of the oil components with the formation of solid substances (X-wax formation); these possess a lower breakdown voltage than the liquid components and therefore initiate the breakdown.

Of the thermal properties, the specific heat and the long-term thermal stability are of particular significance. Since the oil, besides having insulating duties, frequently must also perform as convection coolant, the relatively high specific heat is of advantage. The long-term thermal stability, however, lies at around 90 °C only and so in many devices limits the permissible rated power.

The application of oil as an insulating material occurs almost exclusively in combination with cellulose in the form of paper or pressboard. Oil-impregnated paper, as an electrically extremely strong and under continuous stressing a proven compound dielectric, is the most important high-voltage technical insulating material, without which the present-day concept of many transformers, instrument transformers, bushings, capacitors and cables would be inconceivable.

2.5.2 Paper

Paper for electrotechnical purposes is primarily manufactured from wood-pulp of the slowly growing northern spruce or pine. Only capacitor paper down to about 10 μm thickness is made of rag-pulp. In the first phase of the manufacture the cellulose pulp is separated from the other wood components (e.g. lignin, resin) whereby either acidic or alkaline decomposition takes place in a pulp digester. Cellulose to be used in electrotechnical applications must be carefully washed in order to remove the insulation reducing bleaching agents or acids. The raw cellulose thus obtained (mostly in the form of unbleached lignosulphonates) is dissolved in water and the fibre suspension is then separated into individual fibres and ground. The duration of the grinding and the method essentially determine the paper quality. The watery suspension is then fed to the paper machine which produces machine-fine paper in rolls. By pressure treatment in the glazing rollers, increased smoothness and shine of the paper, as well as greater density, improved breakdown strength and an increased dielectric constant are also achieved.

Cellulose paper is manufactured into transformer paper with thicknesses of 0.05 mm to 0.08 mm, and into cable paper of 0.08 mm to 0.2 mm. Pressboard is made by wet pressing several thin individual layers without bonding material; per mm thickness it consists of 35 individual layers, each about 30 μm thick [*Moser* 1979].

The theoretical density of 1.55 g/cm^3 is not achieved in paper owing to the pore volume of 20...60%; machine-fine paper has 0.65 g/cm^3, high-gloss paper 1.15 g/cm^3 and pressboard reaches 1.3 g/cm^3.

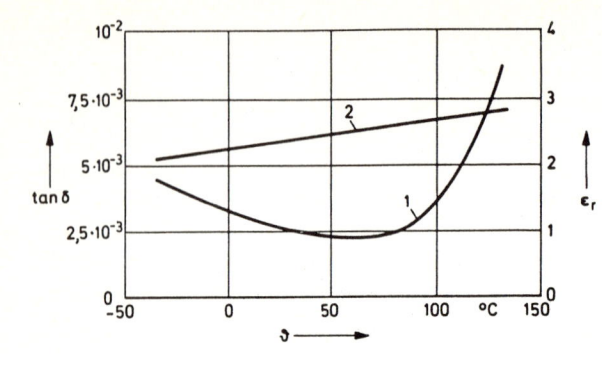

Fig. 2.5-3

Dissipation factor (1) and dielectric constant (2) of paper as a function of temperature

The dielectric constant of cellulose at 20 °C is 5.6, of paper 1.5 to 3.5 and of pressboard 4.5. The dissipation factor lies around $(3...4) \cdot 10^{-3}$, the volume resistivity in the dry state between 10^{15} and 10^{17} Ω cm. The volume resistivity decreases by about a factor of ten for every 1.5% of absorbed water. Fig. 2.5-3 shows the dependence of dissipation factor and dielectric constant upon the temperature.

Paper is very hygroscopic and during storage in an atmosphere of average humidity absorbs 5...10% water. Since the ageing of paper is influenced primarily by water and heat, good drying is of particular importance [*Moser* 1979].

The utilization of paper occurs in the form of hardboard, soft paper and pressboard. Hardboard results on compression with epoxy or phenolic resins and is used for supports, insulating barriers, etc. Soft paper or pressboard are used in the oil-impregnated form in transformers, instrument transformers, bushings, capacitors and oil-filled cables.

2.5.3 Oil-impregnated paper

a) Properties, manufacture

Oil-impregnated paper is the most important compound dielectric for high-voltage insulation systems. Since several layers of paper are usually used, and also on account of the fibrous nature of paper, for the characteristic quantities one may assume a series connection of the insulating materials soft paper and oil. If we consider a plate-type compound dielectric of thickness s and permittivity ϵ, this can be taken as comprising a series connection of a pure oil dielectric (s_1, ϵ_1) and a pure paper dielectric (s_2, ϵ_2). For the resultant dielectric constant we have:

$$\epsilon = \frac{\epsilon_1 \cdot \epsilon_2 (s_1 + s_2)}{\epsilon_1 \cdot s_2 + \epsilon_2 \cdot s_1} \quad \text{with } s_1 + s_2 = s.$$

The unknown thicknesses s_1 and s_2 are eliminated by introducing the fiducial pore volume v:

$$v = 1 - \frac{\gamma_P}{\gamma_Z}.$$

Here, γ_P is the density of the paper containing pores and γ_Z is the density of pure cellulose.

For total impregnation v is equal to the oil volume v_1 and for the volume of paper $v_2 = 1 - v_1$ holds.

2.5 Natural organic insulating materials

Thus we have [*Liebscher, Held* 1968]:

$$\epsilon = \frac{\epsilon_1 \cdot \epsilon_2}{\epsilon_1 + v_1(\epsilon_2 - \epsilon_1)} .$$

Cellulose has a density $\gamma_Z = 1.55 \text{ g/cm}^3$ and $\epsilon_2 = 5.6$. With $\epsilon_1 = 2.2$ for oil, the resultant dielectric constant is

for impregnated cable paper $\qquad \epsilon = 3.1$,
($\gamma_P = 0.75 \text{ g/cm}^3$, $v_1 = 0.516$)

for impregnated capacitor paper $\qquad \epsilon = 4.0$,
($\gamma_P = 1.15 \text{ g/cm}^3$, $v_1 = 0.26$)

for impregnated pressboard $\qquad \epsilon = 4.5$.
($\gamma_P = 1.3 \text{ g/cm}^3$, $v_1 = 0.16$)

With the usual values $\epsilon_1 = 2.2$ and $\epsilon_2 = 5.6$, we have for the ratio of electric stress in the oil/paper lamination:

$$\frac{E_1}{E_2} = \frac{\epsilon_2}{\epsilon_1} \approx 2.55 .$$

The oil is therefore electrically stressed more than the paper; using a finely graded lamination a large number of thin oil films is obtained, the high electric strength of which is responsible for the excellent breakdown field strength of an oil-paper dielectric. The paper promotes the formation of thin oil layers, acts as a barrier to bridging impurities and ensures the mechanical stability of the insulating system.

The breakdown field strengths of oil-paper dielectrics, even for large thicknesses, take values up to 400 kV/mm, and, as operating field strengths in direct voltage capacitors up to 100 kV/mm; during a.c. stressing, up to 20 kV/mm are applied. In a high-quality oil-paper dielectric the dissipation factor is $\tan\delta \approx 3 \cdot 10^{-3}$, the volume resistivity $\rho \approx 10^{15}$ Ω cm, and the permissible temperature limit is ca. 100 °C.

The oil-paper insulation system of an apparatus must be carefully processed during manufacture in order to prevent detrimental gas occlusions, which lead to partial discharges and also reduce the breakdown field strength of oil by the solution of gas. Further, moisture must be wholly removed since it causes a noticeable deterioration not only of the electric strength of the oil, but also of the ageing stability of the paper (see Section 1.4). Processing is carried out by subjecting the paper-insulated live parts in heated vacuum chambers to a vacuum of $10^{-3}...10^{-4}$ mbar and temperatures up to 110 °C; the drying time increases quadratically with the thickness of the insulation and is of the order of days or weeks. The drying procedure is controlled by permanent monitoring of the dissipation factor. The relationship between residual moisture and degassing pressure during drying is described by the adsorption isotherms of Fig. 2.5-4 [*Beyer* 1971]. According to these, a pressure of about 0.1 mbar at 110 °C is necessary to achieve a residual moisture of 10^{-4}.

The dried paper insulation is then impregnated, in vacuum if possible, with recently refined and warm mineral oil. Its hygroscopic nature causes the dry paper to extract the moisture still contained in the oil; in turn, the oil dissolves the residual gases in the paper and so contributes to an improvement of the partial discharge performance.

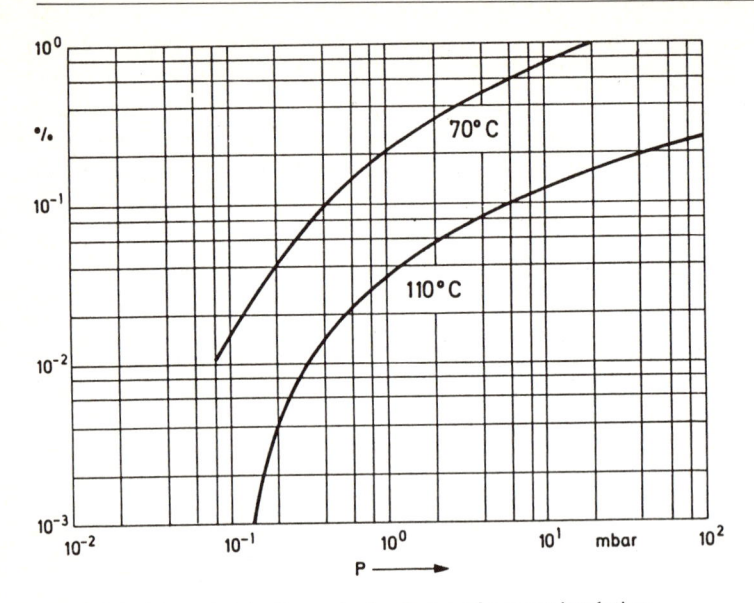

Fig. 2.5-4 Adsorption isotherms in the drying of a paper insulation
p = pressure   v = water content (mass ratio)

b) Oil-impregnated paper as cable insulation[1])

The oil-paper dielectric also plays a prominent role in cable insulation. In the voltage range up to about 60 kV the so-called compound-filled cable was used, which has been practically supplanted by the PE-cable (see Section 2.6.2); for 110 kV and higher the oil-filled cable predominates.

The conductor is first provided with a paper tape insulation wound without overlap 20...30 mm wide and 0.1 to 0.15 mm thick. The paper insulation is dried and impregnated. A low viscosity mineral oil is used for oil-filled cables and a mineral oil thickened with resin additives for compound-filled cables.

In compound-filled cables, the permeating substance is of low viscosity at the impregnating temperature and of high viscosity at ambient and operating temperatures so that bleeding of the cable does not occur during transport and installation. The high viscosity at operating temperature prevents or hinders the permeating substance from running down into the lower-lying cable sections when laid on inclined slopes.

The use of the compound-filled cable, which is of simple construction, is restricted to medium voltages due to the risk of partial discharges. During thermal stress the compound expands more than the lead sheath, which latter then experiences an irreversible expansion. After cooling, gas-filled cavities are created which, with regard to the partial discharge inception, limit the permissible operating field strength to 4 kV/mm.

---

[1]) Summaries in e.g. [*Roth* 1965; *Heinhold* 1965; VDEW-cable handbook 1977; *Ehlers, Lau* 1956; *Lücking* 1981].

2.5 Natural organic insulating materials

Compound-filled cables are in limited use for operating voltages over 60 kV in the form of internal and external gas-pressure cables. In the former, cavity formation in the compound-paper insulation is prevented from doing any harm by keeping the insulation at a gas pressure of 15 bar nitrogen so increasing the inception voltage in the cavities. Then, operating field strengths of 9 kV/mm and, with the addition of SF_6, up to 12...13 kV/mm can be attained. In the external gas-pressure cable, the normal compound-filled cable is placed inside a steel pipe filled with nitrogen (15 bar); the lead sheath acts as a pressure membrane and prevents the formation of cavities or ensures a high pressure in the cavity.

The low-viscosity, mostly inhibited, mineral oils used in oil-filled cables prevent the occurrence of cavities. Expansion vessels are placed at regular intervals of a few km; these keep the cable at a certain pressure. On warming up the low-viscosity oil flows to the expansion vessel without expanding the lead sheath, and from there, on cooling, it is pressed back into the cable insulation once more.

The value of the pressure influences the breakdown field strength as shown in Fig. 2.5-5 [*Roth* 1965]. If the oil pressure is a few bar, one speaks of low pressure oil-filled cables, and, for ca. 15 bar oil pressure, of high pressure oil-filled cables. The operating field strength has values up to 14 kV/mm. In Europe predominantly low pressure oil-filled cables are used.

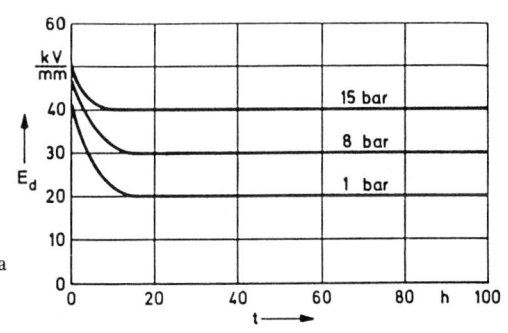

Fig. 2.5-5
Breakdown field strength of oil-filled cables as a function of time at different pressures

With regard to the necessary impulse voltage strength, thinner paper is often arranged in the region of maximum stress, that is, at the inner conductor, whilst in the outer region normal cable paper is used. Besides an increase in the a.c. strength, this kind of lamination results in homogenization of the field distribution, since, due to the higher dielectric constant of the thinner paper layers, the voltage distribution shifts towards the outer regions where the field is weaker. These measures, as well as the use of special less porous paper, make it possible to design a 400 kV oil-filled cable for an impulse level of 1640 kV with an insulation wall thickness of only 28 mm; the maximum field strength at this stress is 93 kV/mm [*Peschke* 1976].

The dissipation factor of oil-filled cables lies around $(2...4) \cdot 10^{-3}$. It can be shown that dielectric losses for increasing transmission voltage limit the transmitting power of a cable [*Peschke* 1973]. Thus, as Fig. 2.5-6 proves, for $\tan \delta = 2 \cdot 10^{-3}$ the transmitting power referred to the conductor diameter reaches a maximum at a transmission voltage of 700 kV. Higher transmission voltages are only practicable if dielectrics with a dissipation factor lower than $2 \cdot 10^{-3}$ are available. Insulation systems of this kind can consist of oil impregnated plastic foils, synthetic paper or paper/plastic combinations.

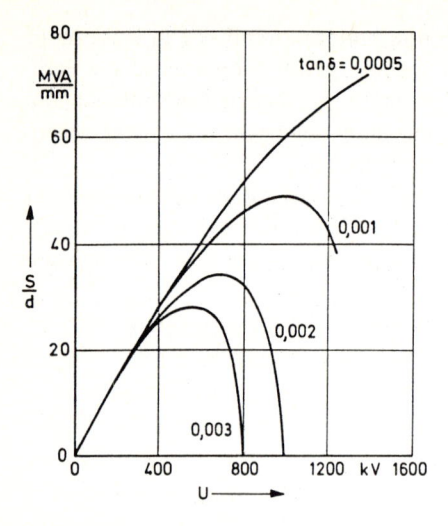

Fig. 2.5-6

Transmitted power S of oil-filled cables referred to the conductor diameter d, as a function of the transmission voltage U, with the dissipation factor as a parameter

2.6 Synthetic organic insulating materials[1])

Synthetic organic insulating materials — commonly classified under the keyword plastics — have acquired a position of great importance in electrical insulation technology as non-hardenable thermoplasts, hardenable duroplasts, elastomers, and a few liquid materials e.g. silicone oils. Their application covers all areas of high-voltage insulation systems.

2.6.1 Molecular configuration and polymerization reactions

Plastics are organic substances consisting of macromolecules. Large molecules have their origin in the ability of the quadrivalent carbon atom to form arbitrarily long chains which can contain some ten thousand atomic groups and have molecular weights of up to a few millions.

The formation of macromolecules occurs by way of a polymerization reaction, during which molecules of low molecular weight (monomers) combine to produce large molecules of high molecular weight (polymers). The reactions in question are polymerization, condensation polymerization and addition polymerization.

During polymerization similar groups of low molecular weight combine without splitting off side products, and, in general, long-chain molecules are formed. The monomers are unsaturated compounds whose double bonds are broken during polymerization leaving free bonds available for chain formation. Polyethylene for example, a typical polymer, is formed from ethylene C_2H_4:

$$\begin{pmatrix} H & H \\ | & | \\ C = C \\ | & | \\ H & H \end{pmatrix}_n \longrightarrow \begin{matrix} H & H & H & H & H \\ | & | & | & | & | \\ -C-C-C-C-C- \\ | & | & | & | & | \\ H & H & H & H & H \end{matrix}$$

---

[1]) Survey in [*Saure* 1979; *Brinkmann* 1975] for example.

2.6 Synthetic organic insulating materials

Other important plastics obtained by polymerization are polyvinylchloride (PVC) and polytetrafluorethylene (PTFE). Common to all the long-chain, sometimes branched, polymers is that they are not cross-linked, i.e. their macromolecules are not chemically bound to one another. This results in the property that these substances, which are hard and tough at low temperatures, show viscous flow at high temperatures because sliding of the unlinked chain molecules against each other is facilitated. These non-cross-linked substances are then designated thermoplasts. The viscous flow at higher temperatures allows their manufacture in injection casting machines or extruders. The raw material, supplied as solid granules, after heating becomes a viscous mass which can be modelled by casting or injection into a mould and removed after hardening; or it can be formed into moulded bodies by extrusion in a press e.g. into a cylindrical cable insulation.

Besides PE, PVC and PTFE, polypropylene, polystyrol and polyamide, for example, also belong to the group of thermoplasts.

Condensation polymerization is the combination of various (dissimilar) groups of low molecular weight to form macromolecules with the splitting off of side-products such as water, ammonia, hydrochloric acid. Examples of polycondensates are polyamides and phenoplasts.

By addition polymerization different types of monomer combine to form macromolecules without splitting off side products. Typical addition polymerization products are epoxy resin (EP) mouldings and polyurethane (PUR).

On account of cross-linkage, condensation and addition polymerization usually lead to spatial chemical bonding of the chain molecules. If the bonding is close meshed, we have duroplasts which, in contrast to thermoplasts, do not exhibit viscous flow, even at higher temperatures, but stay elastic. Thus duroplasts are hardened plastics; they are mainly formed from substances with numerous active groups. Applied as cast-resins they are usually available for processing as a compound system (resin, hardener, accelerator) in liquid form.

In addition to EP-mouldings and PUR, for example, polyester resin, phenolic resin, melamin resin and silicone resin belong to the duroplast group.

When the cross-linkage of the macromolecules is wide meshed, elastomers are produced (e.g. silicone rubber) which at all admissible temperatures show rubbery behaviour and do not exhibit flow. Mechanical strain results in elongation which is reverted on removal of the load.

2.6.2 Polyethylene (PE)

a) Manufacture, properties

Polyethylene belongs to the partly crystalline plastics. During solidification of the melt parallel alignment of some of the chain molecules occurs. The extent of the crystalline region is less in the case of branched polyethylene of low density (LDPE) than in unbranched polyethylene of high density (HDPE).

By chemical or radiation treatment PE molecules can be spatially cross-linked. By detachment of hydrogen atoms bonds for neighbouring macromolecules are set free and then movement of the chains against each another is restricted. In cross-linked polyethylene (XLPE) the thermoplastic properties are forfeited; at higher temperatures, despite melting of the crystallites, no flow occurs, rather a transition to the plastic flow range. The residual strength of XLPE beyond the crystallite melting region and the retention of the structure obtained during cross-linkage are responsible for the thermal overload capacity of XLPE insulations.

PE is manufactured from ethylene either by a high pressure process at 1500 bar and 250 °C, where a branched chain polyethylene of lower density results (LDPE low density polyethylene) or by low pressure solvent polymerization which yields high density polyethylene (HDPE high density PE). Starting from a monomer with relative molecular weight of 28, one obtains polymers with molecular weights up to 500,000.

PE is processed by extruders (tubes, cable insulation) or by injection moulding at temperatures of 200...250 °C. After injecting into the cooler mould, the material shrinks by 3%, does not stick and is therefore easily removed, even complicated pieces. Foils down to 0.01 mm thickness are manufactured by extrusion and stretching, as well as by blowing. The basic construction of an extruder is shown in Fig. 2.6-1 [*Saechtling, Zebrowski* 1971].

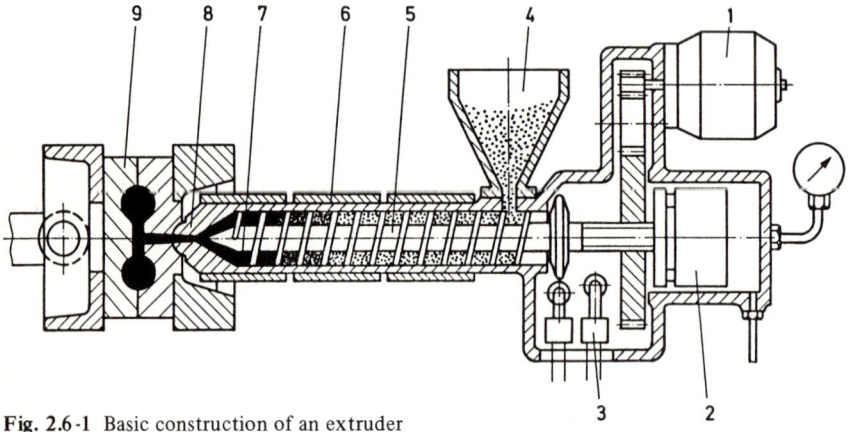

Fig. 2.6-1 Basic construction of an extruder
1 drive for the rotational movement of the helical screw, 2 hydraulic drive for the axial movement of the helical screw, 3 limit switch for the axial helical movement of the screw, 4 hopper for filling, 5 helical screw, 6 injection moulding cylinder with heating elements, 7 plasticized mass, 8 heated injection nozzle, 9 injection moulding tool

PE can be worked by milling, turning, drilling, cutting and punching, and it can be welded. Foils can be joined by applying pressure and heat.

Some important properties of LDPE, HDPE and XLPE are listed in Table A 3.5 of Appendix 3.

PE is viscoelastic with a paraffin-like exterior; even at low temperatures it does not show any appreciable deterioration of properties. The more dense low pressure PE has a greater breakdown field strength and higher dielectric constant than high pressure PE. For foils breakdown field strengths of more than 200 kV/mm are achieved. PE does not contain any polar groups and therefore has a low dielectric constant and a very low dissipation factor. The specific resistance is extraordinarily high and lies around 10^{16} to $5 \cdot 10^{17} \, \Omega \, cm$. However, on account of this, stationary space charges can occur in the material which

2.6 Synthetic organic insulating materials

produce undesirable field variations in consequence. PE can be employed down to −50 °C. It is combustible, its chemical resistance is good except to chlorine, sulphur, nitric acid and phosphoric acid. Under the influence of oxygen the surface becomes brittle.

As a result of its properties PE is applied in high-frequency insulation systems e.g. as full insulation, tape insulation, foam insulation, disc or helical insulation in cables. The most important application in energy technology is as cable insulation.

b) Polyethylene as insulation for high-voltage cables

LDPE and XLPE (on the basis of peroxide cross-linked LDPE) are extensively used for the insulation of cables up to 110 kV. They have almost completely supplanted the compound-filled cable in the medium voltage range. The insulation extruded onto the internal conductor sheath is stressed with a maximum operating field strength of 5 kV/mm. Transition to higher field strengths would make the construction of PE cables for voltages over 110 kV feasible too. The inner and outer conductor sheath is composed of materials mixed with carbon black, which are extruded with the insulation onto the conductor in a single step where possible.

The advantages of PE cables are:

Quick assembly, low weight, small bending radii, installation on slopes or in a vertical position without difficulty.

The main difficulties with the PE and XLPE cables lie in their sensitivity to partial discharges (PD) and the associated question of lifetime. The most minute cavities (microcavities) of 1...30 μm diameter are unavoidable during manufacture, as also are occasional impurities. At these weak spots, under electric stress, partial discharges and development of discharge tracks (trees) can occur which initiate complete breakdown. To avoid these discharges, voltage stabilizers have been developed which are mixed with the insulation during manufacture and either prevent the inception of PD, or make it more difficult, or, if PD damage has already occurred, inhibit the growth of trees; the by-products of cross-linkage in XLPE are effective as stabilizers [*Saure* 1979]. Stabilizers are aromatic materials in most cases.

A special problem in PE cables is the presence of water trees [*Bahder* et al. 1974]. It is known that these begin to grow at sites of high field strength, e.g. at inhomogeneous sites in the conductor sheath, in the presence of water and branch out tree-like without PD occurring. They disappear again on drying. With regard to the creation mechanism of water trees some first steps have meanwhile been taken towards the understanding of this phenomenon [*Heumann* et al. 1980]. However, the extent to which the electrical properties of the cable insulation are affected is still the subject of some dispute; investigations indicate a reduction in the breakdown voltage [*Densley* et al. 1980].

2.6.3 Polyvinylchloride (PVC)

PVC has the structure

```
      H   H   H   H   H   H
      |   |   |   |   |   |
   — C — C — C — C — C — C —
      |   |   |   |   |   |
      H   Cl  H   Cl  H   Cl
```

It is obtained from vinylchloride under high pressure. The C–Cl bond produces a large dipole moment which causes high strength and rigidity of the material as a result of the Van-der-Waals' binding forces.

PVC is often mixed with filler materials (kaolin, quartz) and dyed. It is processed as hard PVC. Softeners can effect a deterioration of the ageing resistance.

Crude PVC is available in powder or granular form and can be processed in extruders or by injection moulding. On account of the chlorine content the processing machines must be hydrochloric acid resistant. Injection moulding processing of cast PVC takes place at pressures up to 600 bar and at a temperature close to the decomposition temperature of 190 °C.

The properties of PVC are summarized in Table A 3.5 of Appendix 3. It can be seen that the electrical and dielectric properties are only average and the poor dissipation factor is especially striking. With the exception of aromatic and chlorinated hydrocarbons, the chemical resistance is good. Hard PVC solidifies at 75...80 °C and soft PVC at −10 °C.

PVC is a cheap insulating material and is used for the core insulation of cables, for example, and for cable sheaths, cable-covers and insulating supports. The application as cable insulation is restricted to about 10 kV on account of the poor dissipation factor.

2.6.4 Polytetrafluorethylene (PTFE)

The chemical structure of PTFE

```
    F   F   F   F   F
    |   |   |   |   |
   -C - C - C - C - C-
    |   |   |   |   |
    F   F   F   F   F
```

is the outcome of polymerization of the gaseous monomers under pressure in water. As in PE, a parallel alignment of some of the long chain molecules with assumption of partly crystalline properties is possible. Above 327 °C PTFE is amorphous; unlike other thermoplasts no melting occurs but a transition to a highly viscous state takes place.

The thermal behaviour, different from other thermoplasts, also demands other processing methods. PTFE is either hot-pressed or sintered after isostatic compression at 350...380 °C. At higher temperatures the pressed parts tend to return to the initial shape. Workpieces must therefore be tempered at temperatures higher than the maximum operating temperature. Machining is possible but requires special techniques. The material can be welded.

The properties of PTFE are also compiled in Table A 3.5 of Appendix 3. The mechanical properties are average, since PTFE is soft and exhibits cold flow. The electrical properties are predominantly good. Tracking strength and arc withstand strength are very good, no conductive residues are produced. On the other hand, the material shows a reduction of the electric strength with time which can probably be attributed to a degree of sensitivity to PD.

PTFE is non-polar; at a density of 2.17 g/cm^3 it has the lowest dielectric constant $\epsilon_r = 2.05$ of all known solid and liquid insulating materials. The dissipation factor, $\tan \delta \approx 10^{-4}$, is very good, and it is particularly noteworthy that at 1 GHz it increases to only $5 \cdot 10^{-4}$.

2.6 Synthetic organic insulating materials

Its thermal stability is also extraordinarily good. In the temperature range $-200...+200$ °C. PTFE shows no appreciable change in properties. It can be continually stressed up to 250 °C. Its chemical resistance to acids and alkalis is remarkable, as also its insolubility in all organic and inorganic solvents. As a consequence of an angle of contact greater than 90° the surface cannot be wetted.

On account of its low loss factor $\epsilon_r \cdot \tan\delta$, even at the highest frequencies, PTFE is especially suitable for high-frequency applications (plugs, tube sockets, bushings). Due to the difficult and therefore expensive processing technique, it has been successfully applied in the area of energy technical insulation problems in only very few cases, e.g. for section isolators in tram and railway overhead supply lines.

2.6.5 Epoxy resin (EP)[1])

Epoxy resins are the most important hard cast type of resins employed in high-voltage insulation technology. Their chemical characteristic is the epoxy group

$$-CH-CH_2 \atop \diagdown \diagup \atop O \;\;,$$

which must be present in sufficient proportion for hardening. Among the large number of different resins, those based on bisphenol A have acquired the greatest significance.

Before we describe the processing, properties and applications a few general definitions shall be presented first.

a) Concepts for cast resins

Cast resins are compound systems consisting of resin, hardener and sometimes accelerator, plasticizer, filler and colouring material. The individual components are stable and can be stored over a certain length of time; the mixture, however, is capable of reaction. There is therefore only a limited time available for the processing of the mixture — the so-called pot life.

The most important concepts are summarised in DIN[2]); an excerpt is quoted below:

Active resins are liquid or liquefiable resins which harden by polymerization or addition polymerization on their own, or with reactive agents, without splitting off volatile components.

Reactive agents are hardeners and accelerators. Hardeners are substances which cause the polymerization or addition polymerization of the resins and so the hardening. Accelerators are materials which speed up the hardening process.

Active bulk resin is a mixture of the above materials ready for processing and sometimes mixed with a filler as well.

Resin castings are hardened substances which are produced by casting in moulds after the active bulk resin is hardened.

---

[1]) Summary in [Saure 1979].
[2]) DIN 16945 Part 1; DIN 16946 Part 1.

b) Processing of EP

Appropriate to the concepts quoted above, processing of EP is done by mixing and homogenizing the individual components — resin, hardener, accelerator and filler — to a resin bulk capable of reaction which is then cast in steel or aluminium moulds. There the hardening to EP-mouldings takes place. The basic design principle of a casting plant is shown in Fig. 2.6-2 [*Kubens, Martin* 1976].

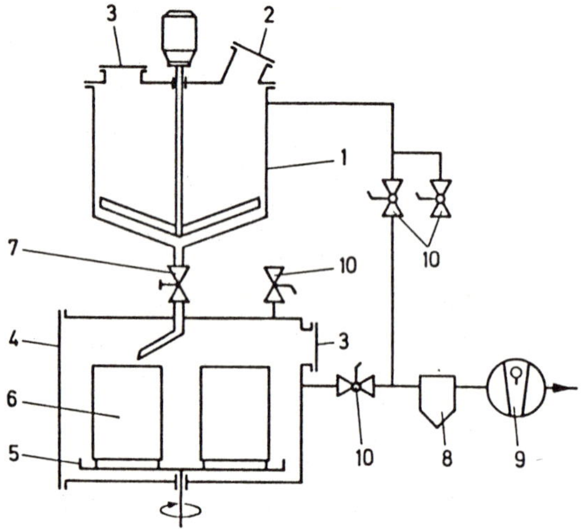

Fig. 2.6-2
Basic design of a casting plant for processing EP mouldings

1 mixer
2 feeder
3 viewing window
4 casting chamber
5 rotating plate
6 casting mould
7 casting valve
8 trap
9 vacuum pump
10 vacuum valve

The adhesion of EP to metal is very good; therefore to ensure easy release from the mould these must be provided with a releasing agent. Casting pieces designed for high electric stress is always carried out in vacuum to avoid cavities.

Hardening, which occurs after gelation, can proceed in two steps: the moulding can be removed after initial hardening at the appropriate hardening temperature; final hardening then occurs in a subsequent heat treatment. However, there are also active bulk resins which harden at room temperatures without heating. But the electrical properties of the (cold-hardened) mouldings, particularly at raised temperatures are usually inferior to those of hot-hardened mouldings. The hardening reaction itself is exothermic. Since the reaction velocity is temperature dependent, removal of the heat of reaction must be taken into consideration.

Pure resins suffer a reaction volume shrinkage during hardening of up to 3%; filled resins, depending on the filler content, of up to about 0.5%. The shrinkage produces internal tension which can lead to crack formation. Moreover, the appreciably larger thermal expansion of resins compared with metals leads to mechanical tension at the metal/resin boundary during a temperature change; this causes cracking risk. This danger is counteracted by the addition of plasticizers which make the mouldings more pliable and ensure casting without crack formation. But resins with plasticizers have, in general, inferior electrical values and a poorer shape retention on heating.

For mass production the mould occupation times play an important role. Longer hardening times support a controlled reaction, but also require longer occupation times. To

2.6 Synthetic organic insulating materials

keep these times short the pressure gelation method was developed, for example, in which the mould temperatures are high [*Dieterle, Schirr* 1972].

Filler materials serve to reduce the volume shrinkage, to increase the compressive strength, to reduce the combustibility and improve the thermal conductivity. Common filler materials, which can constitute up to 65% amount of mass of the resin bulk, are crystalline quartz powder or alumina without water of crystallization.

c) Properties of EP-mouldings

The properties of a few important EP-mouldings are listed in Table A 3.6 of Appendix 3.

It can be seen from the table that the electrical and dielectric properties of EP-mouldings are not excellent. But they are sufficiently good for the good mechanical and thermal properties to be exploited to advantage for high-voltage insulation systems. The easy workability also plays an important role.

EP-mouldings do not react to ether, alcohol, benzene, oil, weak acids and alkalis; they are unstable towards strong acids and alkalis, acetone and chlorinated hydrocarbons. On exposure and UV irradiation leaching occurs and the surface turns yellow. The outdoor stability can be improved by the utilization of cycloaliphatic resins.

d) Application of EP-mouldings

EP-mouldings act as insulation for instrument transformers and dry transformers. Their application is limited by the control of mechanical stresses and the perfect manufacture of bulky insulation systems, where heat dissipation and the volume effect of the electric strength set additional problems.

In indoor installations EP support insulators and EP bushings have wholly supplanted their ceramic counterparts and the support insulators for SF_6 insulated installations are also usually made of EP-moulding. Further important applications are the impregnation of electrical machine insulation, insulating components in high-voltage switches, fibre-reinforced plastic rods for overhead line insulators, pressed parts for mechanical supports, etc.

2.6.6 Polyurethane resin (PUR)

A characteristic of PUR is the urethane group occurring repeatedly

$$-O-\underset{\underset{O}{\|}}{C}-\underset{\underset{H}{|}}{N}-$$

in the molecular chains. PUR is formed by addition polymerization of two liquid reaction components.

Its use as cast resin entails the usual mixing and casting techniques, where quartz powder, kaolin and chalk with an amount of mass of 30-65% are employed as filler materials. The advantage of PUR lies in the simple technology; the mixture cold-hardens rapidly and this results in short mould occupation times.

PUR has a low density, is visco-elastic and can withstand only small mechanical stresses. The electrical properties are comparable with those of EP-moulding (see Table A 3.6 of Appendix 3).

PUR, as a two component varnish, has great significance as a wire varnish. As resin it is used for the insulation of medium voltage instrument transformers, for cable junctions and cable terminations in the medium voltage range, for the impregnation of glass silk and paper tapes for winding insulation, further, for embedding circuits and for casting coils, where casting is also possible without vacuum.

2.6.7 Silicone elastomer

Silicone rubber is produced by vulcanization of silicone caoutchouc, a chain polymer polydiorganosiloxane, by a condensation or addition polymerization mechanism. Fillers are quartz powder, chalk, kieselguhr and titanium dioxide. In the case of hot-hardened elastomers processing is by heating under pressure, for cold-hardened two-component elastomers, by casting.

Moderate mechanical properties contrast with very good electrical properties:

$E_d = 20 \, kV/mm$; $\rho = 10^{15} \, \Omega cm$ (unfilled);
$\epsilon_r \approx 3$; $\tan \delta = 5 \cdot 10^{-3}$; creepage resistance KA3c.

The retention of these properties and the elasticity in the temperature range $-50 \, °C$ to $+180 \, °C$ (long-term thermal stability) is important.

Silicone rubber is resistant to PD, hardly inflammable and, with sufficient filling, self-quenching. It is arc-resistant, since the main chain

$$-\overset{|}{Si}-O-\overset{|}{Si}-O-\overset{|}{Si}-O$$

contains no carbon and disruption of the chain leads to the formation of non-conducting SiO_2.

Silicone rubber is inert to ozone, water repellent (angle of contact $> 90°$), weather-resistant, resistant to weak acids and alkalis, as well as to sulphur and sulphur compounds. It is affected by petrol and aromatic solvents, but less so by aromatic oils.

The main application of silicone rubber is in plug-in type cable terminations in the medium voltage range, as well as in sheds for plastic compound insulators up to the highest operating voltages. Moreover, silicone rubber is employed as conductor and cable insulation, as well as for insulation problems at high temperatures.

2.6.8 Chlorinated diphenyls

Chlorinated diphenyls, also denoted askarels, are made of diphenyl $C_{12}H_{10}$ by replacing 2, 3, 4 or 5 hydrogen atoms with chlorine. Accordingly one differentiates between di-, tri-, tetra- and pentachlorodiphenyls.

The structure of tetrachlorodiphenyl, for example, is:

2.6 Synthetic organic insulating materials

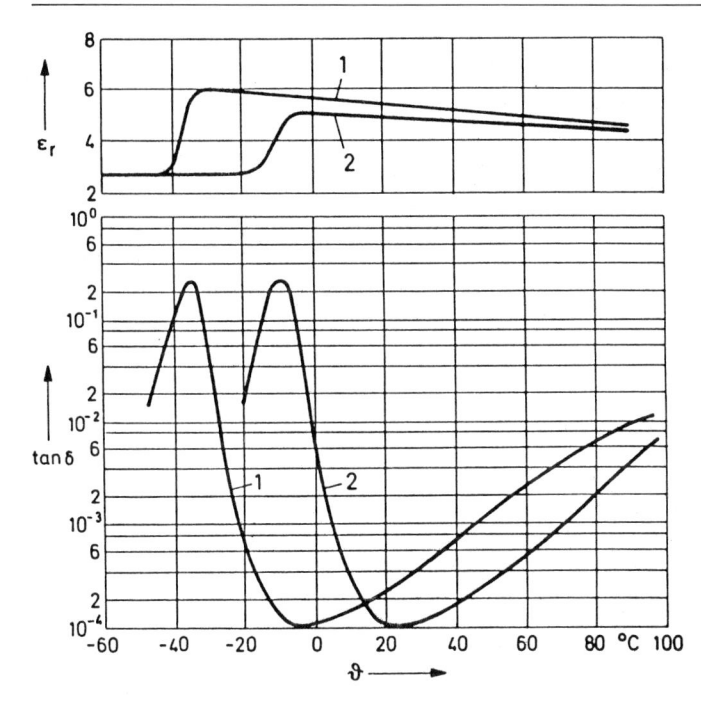

Fig. 2.6-3

Dielectric constant and dissipation factor as a function of temperature for trichlorodiphenyl (1) and pentachlorodiphenyl (2)

Owing to the chlorine content, chlorinated diphenyls have a high density value of $1.4 ... 1.5 \, \text{g/cm}^3$ and even the dielectric constant at 5.5 is, due to the polar configuration, appreciably larger than that of transformer oil. Some properties are summarized in Table A 3.4 of Appendix 3.

The temperature dependence of ϵ_r and $\tan \delta$ is demonstrated in Fig. 2.6-3. Below the solidification point $\epsilon_r = 2.7$ and at the transition point of ϵ_r the dissipation factor has a maximum [*Knust* 1965].

Chlorinated diphenyls are non-combustible because the large affinity of chlorine and hydrogen inhibits the occurrence of free H atoms. Thus the decomposition products of pentachlorodiphenyl in an arc consist of up to 97% hydrogen chloride [*Brinkmann* 1975].

Chlorinated diphenyls have a good resistance to PD, cannot be oxidized and are therefore more or less ageing resistant. The major disadvantage is their toxicity, which makes protective measures during manufacture imperative and permits application in closed systems only. They also react with some plastics.

On account of their high dielectric constant, chlorinated diphenyls are especially suited as impregnants for capacitor dielectrics, because compared with mineral oil it is possible to save 30...40% volume for the same capacitance. In the past chlorinated diphenyls were also used for non-inflammable transformers in public buildings e.g. in theatres. Not least for reasons of environmental protection, these transformers are being replaced by EP-insulated dry transformers. Physiologically safe insulating liquids to replace the askarels are in an advanced stage of development and are being practically utilized.

2.6.9 Silicone oil

Silicone oils represent an alternative to chlorinated diphenyls but they are rather expensive. The main chain consists of silicon and oxygen and organic groups constitute the side-chains. The chemical configuration of methyl polysiloxane for example is:

$$-\underset{\underset{CH_3}{|}}{\overset{\overset{CH_3}{|}}{Si}}-O-\underset{\underset{CH_3}{|}}{\overset{\overset{CH_3}{|}}{Si}}-O-\underset{\underset{CH_3}{|}}{\overset{\overset{CH_3}{|}}{Si}}-O-\underset{\underset{CH_3}{|}}{\overset{\overset{CH_3}{|}}{Si}}-.$$

Chain lengths of up to 800 siloxane units and relative molecular weights up to 60000 can be found.

The most important properties are compiled in Table A 3.4 of Appendix 3. The dissipation factor is independent of frequency and temperature. The high long-term thermal stability at 150 °C is particularly notable. Silicone oils are water resistant and to most chemicals as well as being oxidation resistant, even at higher temperatures. They are soluble in petrol, benzene, ether and alcohols, and they are non-toxic. On thermal dissociation in an arc non-conducting silicon dioxide (quartz) is formed from the main chain.

Irrespective of their high cost, silicone oils can, in principle, be applied as substitute products for mineral oil, e.g. in transformers. The higher admissible working temperature compared with mineral oil allows volume-saving designs. Outdoor insulators under contamination risk often show improved flashover behaviour after siliconizing. A thin layer of silicone fat or silicone oil is a hydrophobic surface and prevents the development of a continuous layer of moisture, even when there is pollution.

3 Design and Manufacture of High-Voltage Equipment

3.1 Structural details in high-voltage technology

The design and construction of high-voltage equipment requires experience in the application of the laws of electric fields. High-voltage technological requirements often hinder a construction which is best from the mechanical and thermal points of view. The job of the constructing engineer is therefore to find the altogether most economical solution. In the following sections a brief introduction will be given to cover the scope and a survey of the established solutions presented.

3.1.1 Basic arrangement of the insulation system

The essential characteristics of an insulation system are the number and type of dielectrics used. Since, as a rule, different potentials have to be insulated against one another as well as a rigid connection made between the electrodes, one cannot manage without solid insulating materials. The boundary surfaces so formed between solid insulating materials and liquid or gaseous dielectrics represent particularly critical regions of an insulation system[1]).

a) Single material configurations

Examples of single material configurations are air clearances in outdoor stations and the insulation of plastic cables. Symmetrical and asymmetrical electrode configurations in general show very different behaviour.

Fig. 3.1-1 shows a qualitative curve of the field strength E along the axis in a symmetrical and an asymmetrical configuration. At the same spacing and electrode curvature symmetrical configurations are a good deal better, since in this case at constant voltage U due to the relationship

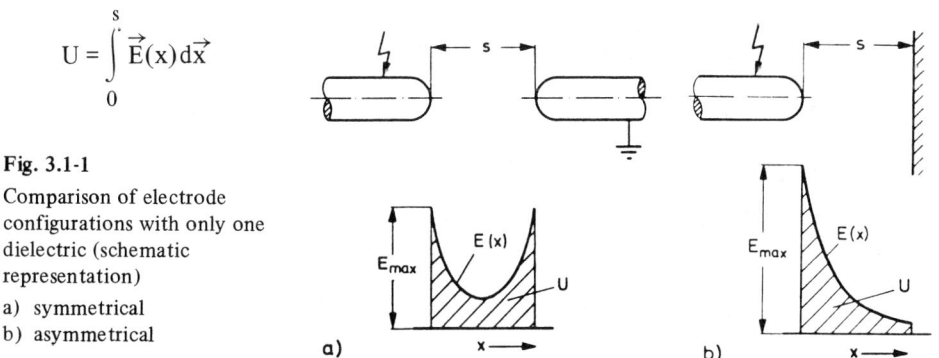

$$U = \int_0^s \vec{E}(x) d\vec{x}$$

Fig. 3.1-1
Comparison of electrode configurations with only one dielectric (schematic representation)
a) symmetrical
b) asymmetrical

---

[1]) Further information in [*Alston* 1968; *Philippow* 1966].

the maximum field strength E_{max} is lower. This can also be expressed with the aid of Schwaiger's utilization factor according to Section 1.1.2:

$$\eta_{symmetrical} > \eta_{asymmetrical} \;.$$

An important application of this knowledge is the possibility of increasing the breakdown voltage of vertical gaps or support insulators for a given spacing s by elevated mounting of the earthed electrode. In Fig. 3.1-2 it is shown that the breakdown voltage of a configuration with strongly inhomogeneous field under impulse voltage increases with increasing length h of the earthed rod from the value for a rod-plate to the value for a rod-rod arrangement; the same measure also results in a reduction of the polarity effect.

Fig. 3.1-2
Symmetrizing a vertical electrode arrangement
a) rod-gap (or support insulator) of spacing s
b) impulse breakdown voltage (qualitative)

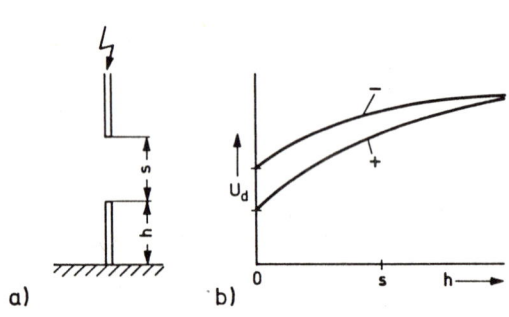

b) Multi-material configurations

In most practical insulation systems several insulating materials are used and boundary surfaces are present between the different dielectrics. The lines of force at these boundaries suffer refraction in such a manner (Fig. 3.1.-3) that the tangential component of the electric field strength remains constant:

$$E_{t1} = E_{t2} \;.$$

Fig. 3.1-3
Electric field strength at the boundary surface of two dielectrics

The normal component is given by the condition of constant dielectric displacement:

$$\epsilon_1 E_{n1} = \epsilon_2 E_{n2} \;.$$

The boundary surface can only be minimally stressed electrically since impurities and humidity cannot usually be avoided and lead to contamination layers. It is therefore an important constructional requirement to keep the field strength at the boundary surfaces low — especially the tangential components. A particularly favourable case is obtained if the boundary surface coincides with an equipotential surface ($E_t = 0$); this is called a "transverse boundary surface". As an example, Fig. 3.1-4 shows the arrangement of insulating barriers in a transformer. During manufacture the barriers are moulded so that they follow the shape of the equipotential surfaces as far as possible.

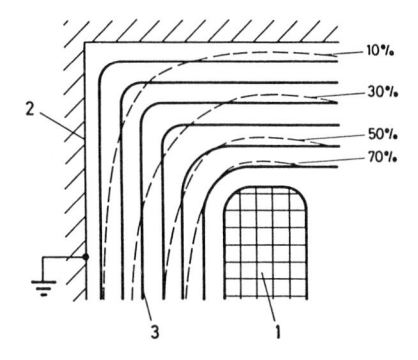

3.1 Structural details in high-voltage technology

Fig. 3.1-4

Insulating barriers with transverse boundary surfaces in a transformer
1 high-voltage winding
2 core
3 insulating barriers

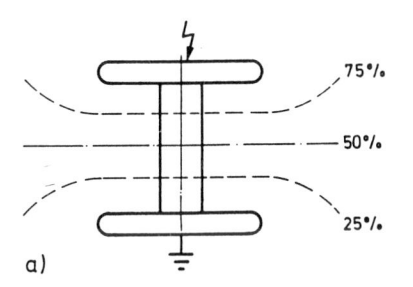

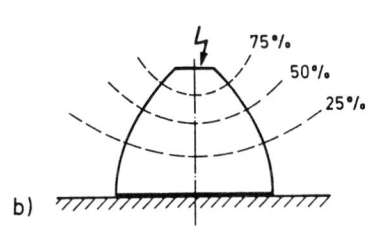

Fig. 3.1-5 Model configurations with longitudinal boundary surfaces
a) homogeneous edge field, b) inhomogeneous edge field

For a "longitudinal boundary surface" the tangential component E_t of the field strength has a finite value whilst the normal component $E_n = 0$. The boundary surface follows a field line, the field distribution is not affected by the solid dielectric introduced. As an example model configurations of support insulators are shown in Fig. 3.1-5; in a) the edge field is homogenized by means of protruding electrodes and in b) the shape of the insulator is adapted to the field distribution.

In technical designs, it is not always possible to prevent the normal components as well as the tangential components of the field strength from having a finite value. This is called an "inclined boundary surface". As an example, consider an insulator with electrodes embedded in a solid insulating material as in Fig. 3.1-6a. This altogether quite favourable configuration can still be appreciably improved by expanding the diameter in the centre of the insulator body as shown by the dotted line, since this would reduce the tangential field strength.

The configuration with electrodes arranged on the surface is particularly unsuitable, as shown in Fig. 3.1-6b. Here, due to relatively high tangential field strengths, surface discharges can scarcely be prevented. The calculation of the inception voltage of this "gliding configuration" is described in Section 1.3.4b.

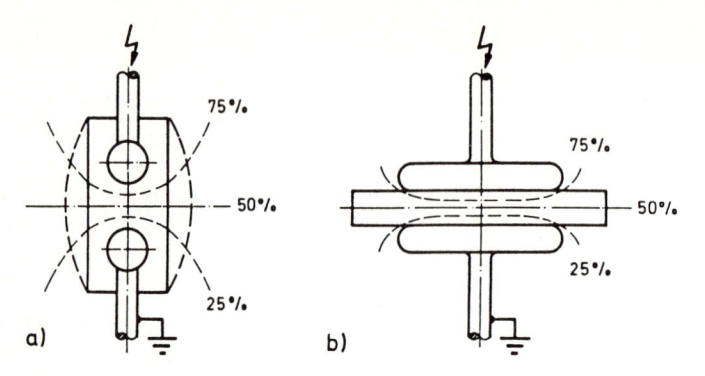

Fig. 3.1-6 Configurations with inclined boundary surfaces
a) embedded electrodes, b) electrodes arranged on the surface

c) Insulating configurations

Where a rigid connection is not established throughout the entire insulation system, as in a solid insulated coaxial cable or in an epoxy resin instrument transformer, special components – the insulators – assume this role. According to Fig. 3.1-7, four types of insulators can be distinguished:

a) support insulators for transmission of compressive and flexural forces,
b) suspension insulators for transmission of tensile forces,
c) bushings for rigid penetration of the electrodes
d) lead-outs for rigid leading out of a voltage-carrying electrode from an earthed region.

In outdoor constructions, the insulators are provided with sheds to increase the creepage path and to prevent the formation of unbroken water channels during rainfall. The shape of the shed depends upon the material used for the manufacture of the insulator and upon the anticipated pollution. As a guiding value for the length of the creepage path, $2...4\,cm/kV$ (referred to the rated voltage) can be assumed. Typical profiles for the sheds

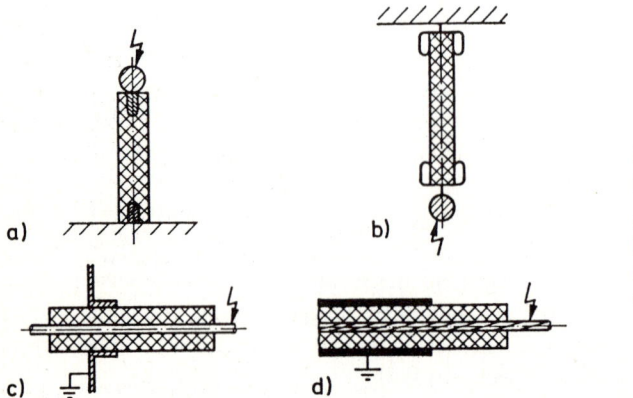

Fig. 3.1-7
Insulating configurations
a) support insulator
b) suspension insulator
c) bushing
d) lead-out
 (e.g. cable termination)

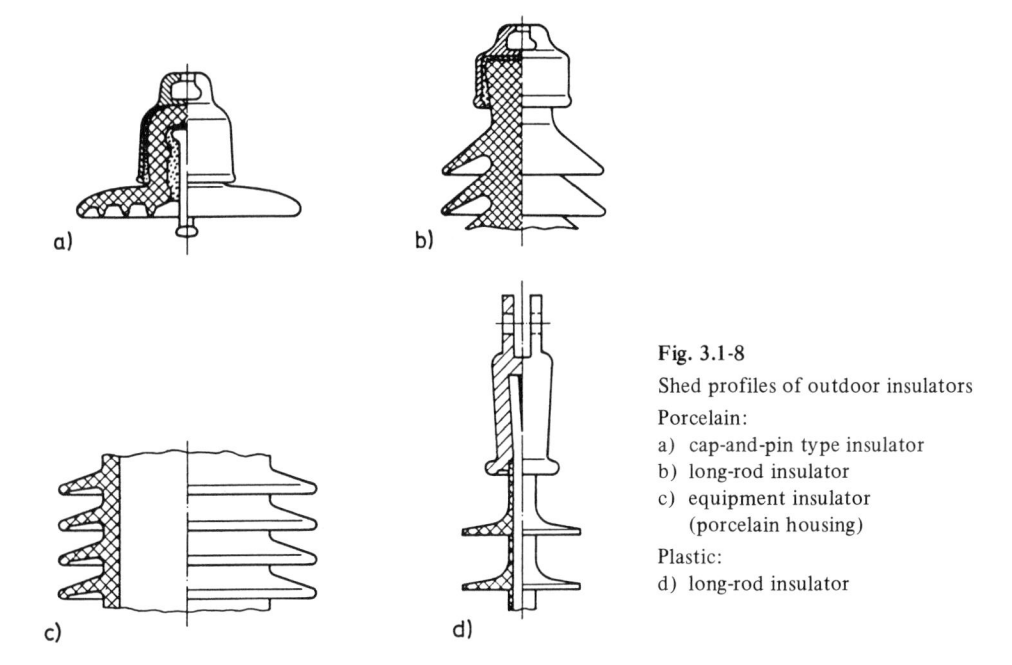

3.1 Structural details in high-voltage technology

Fig. 3.1-8
Shed profiles of outdoor insulators
Porcelain:
a) cap-and-pin type insulator
b) long-rod insulator
c) equipment insulator
   (porcelain housing)
Plastic:
d) long-rod insulator

of porcelain and plastic insulators are presented in Fig. 3.1-8. In plastic insulators the slim shape, especially when hydrophobic shed materials are used, supports a reduction of the specific creepage path without losing the good contamination performance. In gas insulated setups, support insulators are required to maintain the spacing of the leads within the earthed metal housing. In coaxial systems, as in Fig. 3.1-9, disc-type supports (a) or conical shaped ones as on conductor (6) can be used. At rated voltages above 110kV funnel shaped insulators (c) are generally preferred. For three-phase configurations either individual insulators can be used or appropriately complicated insulator shapes must be considered (d). Particularly at high voltages the shape of support insulators for gas-insulated systems must be chosen with due regard to the curvature of the electric field at the boundary surface.

3.1.2 Measures to avoid intensification of electric stress

Various types of insulating materials and the boundary surfaces between them are encountered in technical insulation systems. Uniform stressing results if each part of the insulation system under voltage is stressed by about the same proportion of the permissible value. In the design of an insulation system the stress distribution can be influenced by the arrangement and shape of the electrodes and the insulating body. Where critical points such as electrically stressed boundary surfaces cannot be avoided anyway, one must attempt to limit their stress to permissible values.

a) External screening electrodes for field control

If the dimensions of the electrodes in insulation arrangements are small compared with the gap length, the field strength is high in the neighbourhood of these electrodes. In

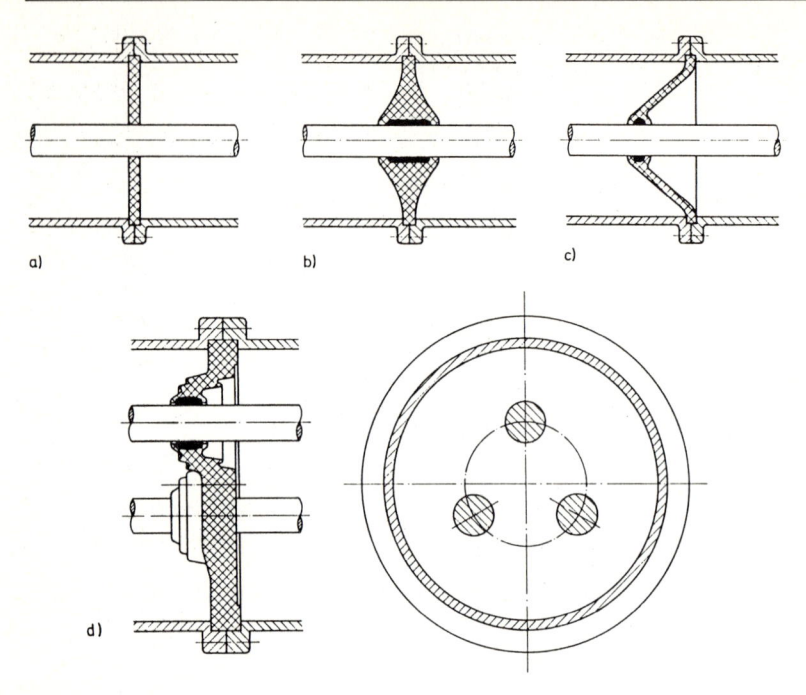

Fig. 3.1-9 Coaxial support insulators for metal encapsulated gas insulated systems
a) disc-type support insulator, b) conical support insulator, c) funnel type support insulator,
d) funnel type support insulator for 3-pole encapsulation (Siemens)

these cases, using external screening electrodes, substantial homogenization of the stress can be achieved by field control.

An example for external screening electrodes are the top-electrodes for cylindrical insulating arrangements in air such as support insulators, high-voltage capacitors or testing transformers. Fig. 3.1-10 shows various designs, where the points of highest field strength are indicated by arrows. The mounted hemisphere as in a) is practically useless; even for

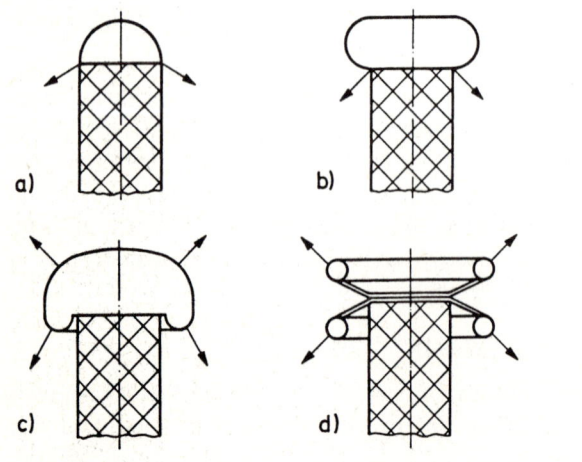

Fig. 3.1-10
Shapes of external screening electrodes
a) and b)   bad
c) and d)   good

3.1 Structural details in high-voltage technology

the protruding electrode mounted as in b) the critical boundary surfaces lie in the region of high field strength and here gliding discharges can occur. In the better design of a protruding electrode as in c) the critical boundary surfaces are in field shadow. Another good shielding electrode, the double toroid shown in d), should be mentioned. This electrode shape has proved particularly successful for very high voltages because of convenient connecting possibilities, since the junction point of an inserted conductor can be arranged to lie in field shadow.

To estimate the spherical and cylindrical electrodes, one starts with the field produced by concentric spheres with internal radius r_k and external radius R. In a rough approximation R corresponds to the clearance of the earthed parts. The field strength at the inner sphere at voltage U is:

$$E_{max} = \frac{U/r_k}{1 - r_k/R} .$$

The highest possible field strength is the breakdown field strength E_d. For $R \gg r$ as an approximation for the inception voltage we have:

$$U_e \approx E_d r_k .$$

With air as the dielectric and alternating voltages the effective value of E_d can under favourable conditions be assumed to be about 20 kV/cm. For the diameter $d = 2 r_k$ the following useful thumb-rule is obtained:

$$d \text{ in mm} \geqslant U_{eff} \text{ in kV} .$$

This relationship is a useful approximation for smooth sphere-like electrodes up to voltages of about 1 MV. At still higher voltages the top-electrode diameter must increase overproportionally and one can finally rely upon permissible field strengths of only about 10 kV/cm. Top-electrodes of this size with smooth surfaces can only be realized with difficulty and at high cost. An alternative, economical at very high voltages and also very efficacious, is the subdivision into electrode elements. Here, as shown in Fig. 3.1-11, numerous individual partial electrodes are mounted on a shell such that the effect of a top-electrode with a large diameter is achieved.

Cylindrical electrodes are mainly required as connecting tubes in testing setups and switching stations. In contrast to concentric spheres, the inception voltage for coaxial

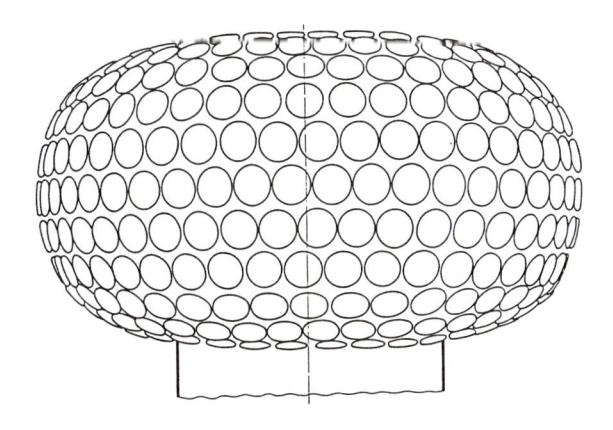

Fig. 3.1-11
Top-electrode made up of electrode elements

cylinders has no finite limiting value for large external radius R. With internal radius r_z, we have for cylinders:

$$U_e = E_d \, r_z \ln \frac{R}{r_z}.$$

On equating this with the corresponding relationship for spheres, under the assumption of the same E_d-values, for large R one obtains:

$$\frac{r_k}{r_z} = \ln \frac{R}{r_z}.$$

If it is assumed, for estimation purposes, that $R > 20\, r_z$, it follows that $r_k > 3\, r_z$. Thus, for the same inception voltages, cylinders need only have about 1/3 the diameter of spheres.

In complicated configurations approximation by spherical or cylindrical arrangements is only conditionally possible. By numerical field calculation, however, even in these cases exact dimensioning can be undertaken [*Moeller* et al. 1972; *Feser* 1975].

b) Internal screening electrodes for field control

The field concentration at critical boundary surfaces can also be relieved by means of protruding internal electrodes. As an example the 10 kV epoxy resin support insulator shown in Fig. 3.1-12 shall be described. By embedding the fittings, the field strength along the surface of the insulator is made considerably homogeneous. Compared with externally cemented support insulators, the constructional height can then be notably less.

The coaxial support insulators reproduced in Fig. 3.1-9 are another example; the protruding electrodes reduce the field strength at the boundary surface in the critical region near the inner conductor.

Protruding electrodes are also usual in the gap part of oil-insulated equipment. Fig. 3.1-13 shows the design sketch of an oil insulated testing transformer with internal ring electrodes which, when properly arranged, effect an increase of the external flashover voltage. These are provided with an insulating bandage of cable paper to increase the permissible field strength at the electrode.

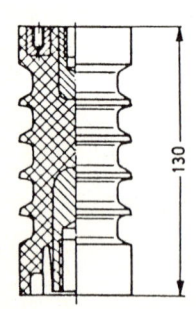

Fig. 3.1-12 10 kV support insulator made of epoxy resin for indoor application

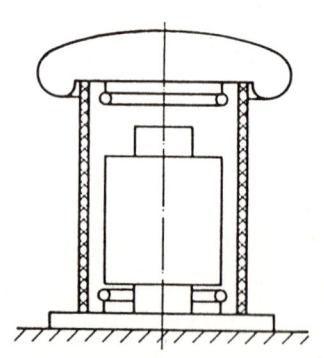

Fig. 3.1-13 Internal screening electrodes in oil insulated high-voltage equipment in the insulated housing construction

3.1 Structural details in high-voltage technology

c) Potential control by intermediate electrodes

A further means for the homogenization of electric stress is the insertion of intermediate electrodes which cause certain equipotential surfaces to be set up in the field space. As a rule, the subdivision of the arrangement aims to yield approximately equal partial voltages. Potential control of the intermediate electrodes for direct voltage is achieved through the effective partial resistances, for alternating and impulse voltage through the effective partial capacitances. The more common second case shall be treated here. Capacitive potential control is achieved by the partial capacitances of the arrangement (self-control), and in addition externally arranged grading capacitors can also be provided (external control). If the capacitances to only one of the external electrodes play a role, this is called simple coupling. On the other hand Fig. 3.1-14 shows a doubly coupled capacitor chain. For identical values of the main capacitances C_ν, the coupling capacitances $C_{h\nu}$ and $C_{e\nu}$ lead to an unequal voltage distribution. For different values of C_ν the coupling capacitances should be so graded that equal partial voltages ΔU_ν result.

The voltage distribution for given elements can be calculated easily using the methods of numerical network calculation and existing computer programmes; the real problem in this case lies in the determination of the partial capacitances of the equivalent circuit. For simple coupling the calculation can even be done manually and one may also choose, to advantage, the straightforward graphical grid method [*Philippow* 1966].

For double coupling an approximate solution can be given for the case that the coupling capacitors are small compared with the main capacitances. The voltage distribution is determined separately for each coupling and then the two are superposed. Fig. 3.1-15 shows the voltage distribution along the chain where ν is the step number out of a total number of n steps. The ideal voltage distribution would be linear (curve 1). If only earth capacitances are present, it would curve downwards (curve 2); if only high-voltage coupling capacitances are present, it would curve upwards (curve 3). The resultant approximation (curve 4), taking both couplings into account is determined by superposition. To do this, curve 3 is lowered by the amount ΔU by which curve 2 deviates from the linear distribution (curve 1). It is evident that the voltage distribution is linearized by the effect of the coupling capacitances on the high-voltage side.

Finally, the possibility of total compensation of the earth capacitances by arranging additional grading capacitors shall be mentioned. Fig. 3.1-16 shows the ν-th junction of

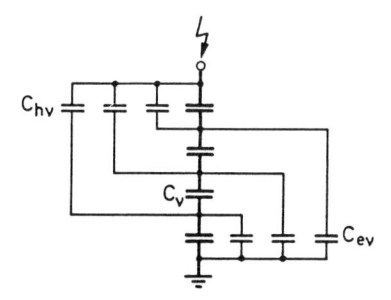

Fig. 3.1-14 Capacitor chain with double coupling

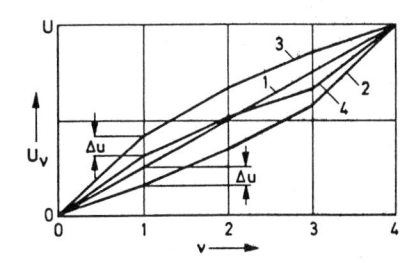

Fig. 3.1-15 Approximate determination of the voltage distribution for double coupling

an n-element doubly coupled capacitor chain. For given $C_{e\nu}$ and constant C_ν, $C_{h\nu}$ should be calculated so that the voltage distribution becomes linear. The voltage drops on the main capacitances C_ν should all be equal to ΔU and add up to the externally applied voltage $U = n\,\Delta U$. It follows from this that $i_\nu = i_{\nu+1}$ must be satisfied. The current law for the junction then gives:

$$(U - U_\nu)\omega C_{h\nu} = U_\nu \omega C_{e\nu}.$$

With $U_\nu = \nu\,\Delta U$ one finally obtains:

$$C_{h\nu} = C_{e\nu}\frac{\nu}{n-\nu}.$$

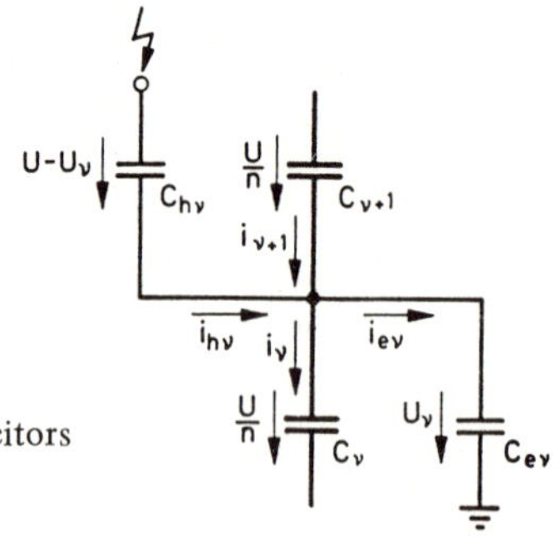

Fig. 3.1-16 Design of grading capacitors for capacitor chains

With the earth capacitances $C_{e\nu}$ known, either by measurement or by calculation, it is possible to calculate the necessary coupling capacitances to the high-voltage electrode $C_{h\nu}$.

3.1.3 Rigid and leak-proof connections to insulating parts

Construction of high-voltage equipment without the use of solid insulating materials is practically impossible. It is then that the problem of rigid connections between metal and insulated parts arises.

Connections which transmit force through a point, such as screws or rivets, are generally unsuitable for insulating materials. In epoxy resin moulding, the metal fittings can be cast (Fig. 3.1-12), whereas in porcelain, fittings will have to be attached with Portland cement or lead (Fig. 3.1-18). In both these solutions the different thermal expansion of the materials involved must be taken into account. To achieve good bonding, the surfaces involved are often provided with a rough outer layer. Embedded parts can be reverse-cut for improvement of the rigid connection.

Rigid connections to insulating parts are often also seals that need to be made leak-proof. As gasket material synthetic oil-resistant caoutchouc is predominantly used which, compared with normal rubber, becomes less brittle after ultraviolet irradiation.

When leak-proof seals are built, one should remember that gaskets tend to be elastically deformed by the effect of forces, and are incompressible like liquids. However, the deformation must be restricted to prevent cracking of the gasket material. As an important scaling factor a necessary surface pressure of $2...3\,\text{N/mm}^2$ on the seal area can be assumed; the size of the seal is of far less significance than the quality of the actual sealing site.

Examples for the application of sealing rings are shown in Fig. 3.1-17. The groove prevents the deformation of the gasket beyond permissible limits. Sealing rings, especially as so-called O-rings, have proved extremely useful.

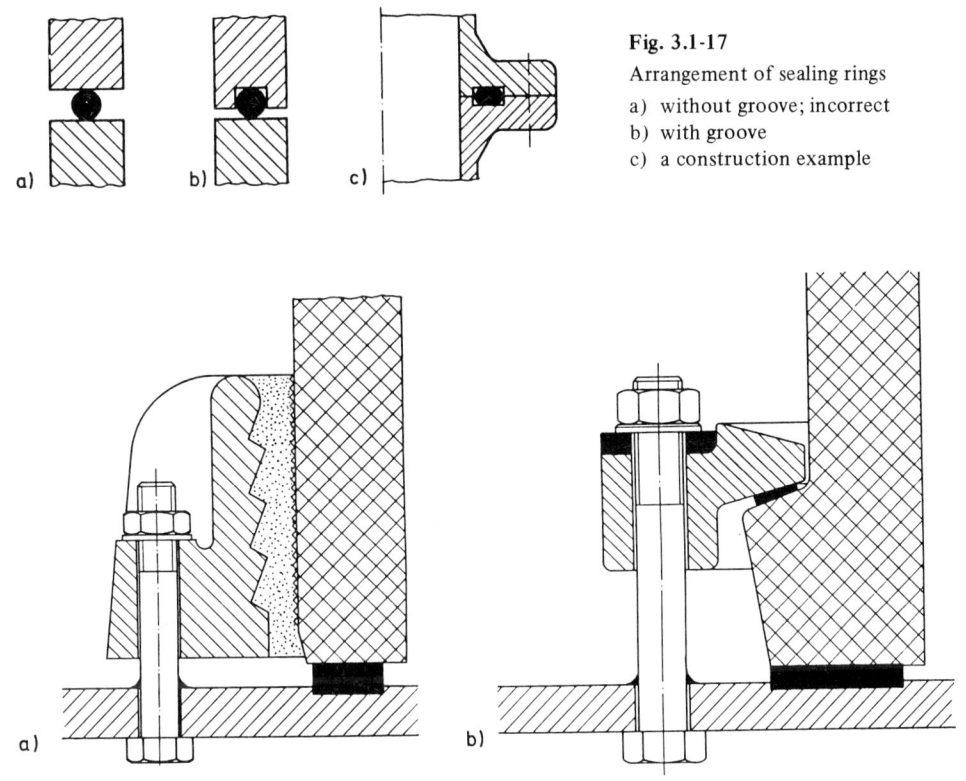

3.1 Structural details in high-voltage technology

Fig. 3.1-17
Arrangement of sealing rings
a) without groove; incorrect
b) with groove
c) a construction example

Fig. 3.1-18 Fixing porcelain housing over flat gaskets
a) cemented flange, b) cam-mounting

For seals between metal and insulating parts, O-rings can quite often not be used since the required groove is not always feasible. Fixing porcelain on a metal component with the aid of a cemented flange and with flat gaskets is shown in Fig. 3.1-18a. When a cam-mounting is used according to Fig. 3.1-18b, special attention must be paid to ensure that the socket holder on the porcelain is subjected preferably to compression only, rather than bending under the action of the compressing force. Point loading of the porcelain surface should be avoided by underlying the cam. If the porcelain housing is ground, greater dimensional stability can be achieved; on the other hand when the porcelain is unground and glazed it shows distinctly higher mechanical strength.

An example for tight and rigid attachment of hardboard tubes is shown in Fig. 3.1-19. The split metal flange 1 which at the same time is formed as a screening electrode, is pressed by a screw joint radially against the insulating tube 2 which is supported on the inside by a steel cylinder 3. The flat gasket 4 is compressed by tightening the screws 5 and seals off the entire front face.

The designs reproduced above can only give some indication of the numerous possibilities. Correct dimensioning, good finish and careful checking of the seals are essential provisos for the reliability of oil-insulated and gas-insulated high-voltage devices.

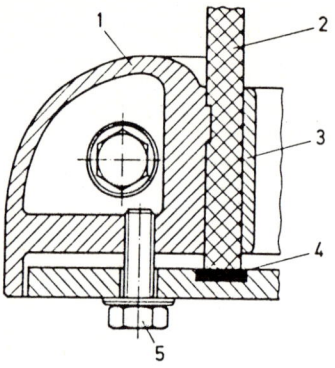

Fig. 3.1-19 Fixing an insulating cylinder made of hardboard
1 metal flange, 2 insulating tube, 3 steel cylinder, 4 flat gasket, 5 screws

Fig. 3.1-20 Oil expansion system for power transformers
1 expansion chamber, 2 desiccator

3.1.4 Measures for air sealing oil-insulated devices

Transformer oil which is used in many high-voltage devices shows a temperature dependent volume change of 7...8% for every 100 °C. Such large variations of working temperature are not at all unusual for outdoor equipment. One is therefore compelled to take this change of volume into account during construction [*Lutz* 1960]. In so doing it must be remembered that the insulating oil, if in contact with the surrounding air, absorbs oxygen and moisture, which results in accelerated ageing. The water content of oil-insulated devices without air sealing is increased to a high degree by condensation on the container walls.

In devices with a large volume of oil perfect air sealing is not possible. Fig. 3.1-20 shows a solution typical for power transformers, namely with an expansion chamber 1. This has the advantage that the exchange surface with air is small and the oil there has approximately the low temperature of the surrounding air, with the result that the exchange mechanisms are retarded. With the addition of a desiccator 2 moisture is also kept away from the oil. The desiccator contains a strongly hygroscopic salt (silica gel) which turns from blue to red when it absorbs water.

In devices with an oil volume of up to a few 100 ℓ perfect air sealing can be achieved which also keeps atmospheric oxygen away from the insulating oil. The changes of volume can at the same time be controlled by systems which have proved successful in various ranges of application.

Small devices with metal encapsulation can be constructed as in Fig. 3.1-21a with an elastic housing. When the volume increases as a result of a rise in temperature, the side walls of the sheet metal housing are deformed as a membrane would be. At the same time the internal pressure remains almost constant. This type of construction predominates in power capacitors.

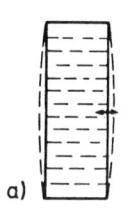

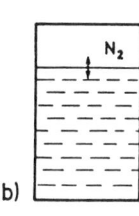

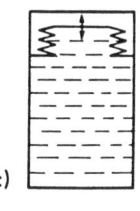

3.1 Structural details in high-voltage technology

Fig. 3.1-21
Perfect air impermeability in oil insulated equipment
a) elastic housing
b) nitrogen cushion
c) bellows

Another solution is a nitrogen cushion to absorb the volume changes of the oil, as in Fig. 3.1-21b. With an increase in temperature, the pressure within increases, the oil expands and the nitrogen gas is compressed. The system therefore works under variable pressure. If the pressure increases, then according to Henry's law, more nitrogen gas dissolves until saturation pressure is reached. On rapid cooling super-saturation results which, under specially unfavourable circumstances, can lead to gas release in the form of bubbles which greatly reduces the electric strength. This can certainly be prevented in equipment with oil at rest, as in instrument transformers and high-voltage capacitors, by the right choice of dimensions for the nitrogen cushion [*Kind* 1959].

The volume change of the oil can be controlled equally well by introducing metal bellows as in Fig. 3.1-21 c. As in the case of elastic housing, the oil here is under constant pressure. A disadvantage is the relatively high cost of bellows; yet the installation position of the equipment (vertical or horizontal) is arbitrary for this solution. It is preferably applied in instrument transformers and high-voltage capacitors. Bellows can also be made of plastic materials, but here the diffusion of oxygen and water vapour must be taken into account.

3.1.5 Temperature rise calculation of insulation systems

Electrical insulating materials as a rule are poor thermal conductors; they can therefore dissipate the heat loss in high-voltage conductors only when large temperature differences exist. However, since the maximum permissible temperatures are relatively low, as discussed in Section 2.2.2, the temperature rise calculation is of great practical significance to the design of high-voltage insulation systems and equipment. A few simple examples shall be given here as an introduction.

Example 1: Adiabatic heating

For a copper busbar (length l, area of cross-section A), calculate the final temperature ϑ_e in the short-circuit case, given that the total switching-off time is 1 s, the current density $S = 120$ A/mm^2 and the initial temperature $\vartheta_a = 60$ °C.

$$\Delta\vartheta = \frac{W}{c \cdot m} = \frac{R I^2 t}{c l A \gamma} = \frac{\rho}{c \gamma} S^2 t \,.$$

For copper, we have:

$$\rho = 0.0178 \, \frac{\Omega \, mm^2}{m} \,, \quad c = 380 \, \frac{Ws}{kg\,K} \,, \quad \gamma = 8920 \, \frac{kg}{m^3} \,.$$

It follows that:

$$\vartheta_e = \vartheta_a + \Delta\vartheta = 60 \,°C + 76 \,°C = 136 \,°C \,.$$

Example 2: Thermal conduction

A high-voltage cable with an oil-paper insulation of length l is laid underground at a depth $h = 70$ cm and has specific losses $P'_v = 40$ W/m. The outer diameter of the cable is $D = 15$ cm, the conductor diameter $d = 3$ cm. How high is the conductor temperature ϑ_L at an ambient temperature $\vartheta_U = 20\,°C$?

The temperature difference $\Delta\vartheta$ between the conductor temperature and the ambient temperature depends upon the total losses P'_v of the cable which must be dissipated radially outwards, and also upon the sum of the thermal resistances ΣR_t. The thermal current follows a rule similar to Ohm's law [*Happoldt, Oeding* 1978]:

$$\vartheta_L - \vartheta_U = \Delta\vartheta = P'_v \, l \, \Sigma R_t \, .$$

The specific thermal conductivity of the oil-paper insulation is $\lambda_1 \approx 0.2$ W/m K, and that of soil $\lambda_2 \approx 1$ W/m K. For concentric layers, the thermal resistance R_{t1} of the insulation can be calculated from this according to the formula

$$R_{t1} = \frac{1}{\lambda_1 \cdot 2\pi \, 1} \ln \frac{D}{d} \, .$$

For the soil, we have:

$$R_{t2} = \frac{1}{\lambda_2 \cdot 2\pi \, 1} \ln \frac{4h}{D} \, .$$

Thus we obtain:

$$\Delta\vartheta = P'_v \, l \, (R_{t1} + R_{t2}) = 40 \, \frac{W}{m} \left(1{,}28 \, \frac{mK}{W} + 0.47 \, \frac{mK}{W} \right) = 70\,°C$$

$$\vartheta_L = \vartheta_U + \Delta\vartheta = 20\,°C + 70\,°C = 90\,°C \, .$$

Example 3: Convection

A switch-box with heat dissipating surface $A = 4\,m^2$ is provided with a plug-in panel which, for a uniform distribution of the heat sources, causes a power loss $P_{in} = 450$ W. Investigate whether at room temperature $\vartheta_u = 30\,°C$ the permissible box temperature ϑ_s of $60\,°C$ is maintained [BBC 1977]. We have:

$$P_{out} = \alpha \cdot A \cdot (\vartheta_s - \vartheta_u) \, .$$

With $\alpha = 4\,W/m^2\,K$, we have:

$$P_{out} = 4 \, \frac{W}{m^2 K} \cdot 4\,m^2 \cdot 30\,K = 480\,W > P_{in} \, .$$

The permissible temperature in the box is therefore not exceeded.

Comprehensive thermal calculations, for example on cylindrical windings of transformers in oil, are possible on the basis of more advanced literature [*Richter* 1963; VDI 1974; *van Leyen* 1971].

3.2 High-voltage capacitors[1])

Capacitors are particularly frequently used in high-voltage circuits. Even for very high voltages they can be built almost ideally i.e. low loss and without inconvenient self-inductance. In high-voltage networks they are used as

Power capacitors	for shunt compensation during inductive load,
Series capacitors	for series compensation, especially in long transmission lines,
Coupling capacitors	for carrier frequency communication setups,
Divider capacitors	for capacitive voltage transformers,
Filter capacitors	for converter stations, especially in high-voltage d.c. (HVDC) transmission systems,
Grading capacitors	for voltage distribution in circuit breakers,

in laboratories as

Smoothing capacitors	in the generation of high direct voltages,
Impulse capacitors	in the generation of high voltage and current pulses,
Measuring capacitors	for voltage and dissipation factor measurement.

In addition, special high-frequency capacitors are used in oscillator circuits, for transmitters for example.

A survey of the most important types and their construction shall be given in this section.

3.2.1 Basic configurations

High-voltage capacitors are always designed with as homogeneous a field as possible. For the ideal parallel-plate capacitor of capacitance C, after Fig. 3.2-1, neglecting the edge-field and with an alternating voltage U of angular frequency ω, we have for the power:

$$P = U^2 \omega C = \left(\frac{U}{s}\right)^2 \omega \epsilon_0 \epsilon_r A s$$

and for the power density:

$$P' = E^2 \omega \epsilon_0 \epsilon_r .$$

In an analogous manner, for direct voltage U we have the energy:

$$W = \frac{1}{2} U^2 C = \frac{1}{2} \left(\frac{U}{s}\right)^2 \epsilon_0 \epsilon_r A s$$

and the energy density:

$$W' = \frac{1}{2} E^2 \epsilon_0 \epsilon_r .$$

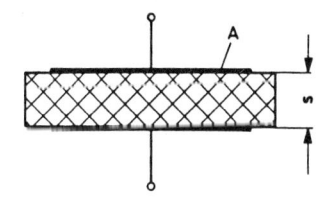

Fig. 3.2-1 Parallel-plate capacitor
A = electrode area
s = thickness of the dielectric

---

[1]) Comprehensive treatment in [*Sirotinski* 1958; *Alston* 1968; *Liebscher, Held* 1968].
The most important properties of high-voltage capacitors are defined in [VDE 0560] as well as in [IEC Publication 70].

High power and energy densities are therefore obtained for a high dielectric constant ϵ_r, but particularly for a high operating field strength,

$$E = \frac{U}{s}.$$

The simple parallel-plate capacitor after Fig. 3.2-1 is a typical gliding configuration and even for relatively low voltage discharges would occur at the electrode edges. For the inception voltage U_e under alternating voltage we have, after Section 1.3-4b, approximately

$$U_e \sim \sqrt{s}.$$

The corresponding field strength in the dielectric is:

$$E_e = \frac{U_e}{s} \sim \frac{1}{\sqrt{s}}.$$

The operating field strength E must naturally lie below the inception field strength E_e, so that damage of the dielectric is avoided. A high operating field strength can be attained if the edge field is controlled by appropriate shaping of the electrode edges or by the choice of a very thin dielectric. However, this has a low operating voltage of the parallel-plate capacitor as consequence, as may be seen from Fig. 3.2-2.

An example of the application of capacitance grading are ceramic capacitors with protruding edges, as they are designed for high-frequency circuits for voltages up to a few 10 kV [*Zinke* 1965; *Liebscher, Held* 1968] (Fig. 3.2-3). Special compounds of titanium dioxide with $\epsilon_r = 30...80$ are usually used as dielectric.

The dissipation factor of these materials decreases with increasing frequency and reaches values of $\tan\delta < 10^{-3}$ at frequencies above 1 MHz; one can expect $E_d = 100...200 \, kV/cm$ for the breakdown field strength. The electrode coatings are applied by metal spraying methods. The connecting tabs are soft-soldered. In this type of construction the edge field contributes a considerable part to the effective capacitance.

If large capacitances are to be realized at high voltages, the only means of obtaining high operating field strengths is by choosing very low dielectric thickness. Dielectric and electrodes are then in thin foil form, so that they can be wound to achieve a technically

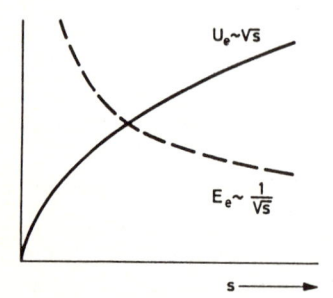

Fig. 3.2-2 Inception voltage and inception field strength of a parallel-plate capacitor without capacitance grading

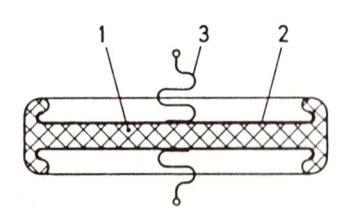

Fig. 3.2-3 Section through a ceramic capacitor with capacitance grading
1 ceramic, 2 electrode, 3 connecting tabs

3.2 High-voltage capacitors

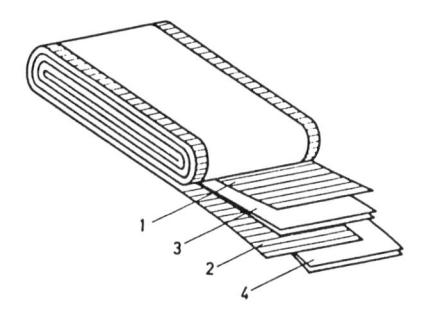

Fig. 3.2-4 Basic construction of a wound capacitor (flat winding)
1, 2 electrode foil, 3, 4 dielectric foil

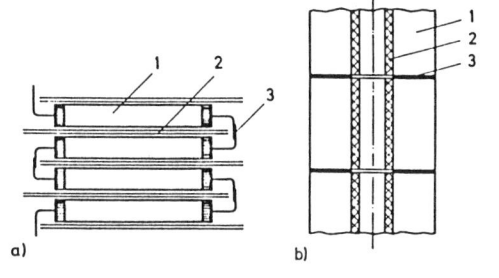

Fig. 3.2-5 External series connection of capacitor windings
a) flat winding, b) cylindrical winding
1 winding, 2 insulating material, 3 electrical connection

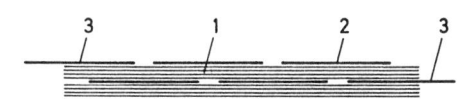

Fig. 3.2-6
Internal series connection of capacitor windings
1 dielectric, 2 intermediate electrode, 3 connecting terminals

suitable shape. Fig. 3.2-4 shows the principle of a wound capacitor whose electrodes are usually made of about 10 μm thick aluminium foil. Several layers of impregnated paper or of plastic films also find application as dielectrics. Due to the winding, each electrode is used on both sides; the effective area of an individual winding can come to a few m². The external connections are made such that the electrode foils project on either side, as shown in the figure, or by inserting connecting tabs.

Capacitor windings are manufactured by winding on a cylindrical mandrel. One distinguishes between flat and cylindrical windings depending on whether the winding, after removal from the mandrel, is pressed flat or retains its cylindrical shape. The final shape is obtained by pressing the winding or by hardening the impregnating or casting medium. Since the voltage of a single-turn capacitor is limited because of the necessary small thickness of the dielectric, a series connection of a large number of windings must always be made in a high-voltage capacitor.

Flat windings are stacked one above the other as in Fig. 3.2-5a and are kept under compression. For series connection of the windings galvanized copper foils, inserted either between windings or connected to the overlapping electrode foils, are externally soldered. Cylindrical windings can be wound quite tightly because of the insulating tube which remains inside the winding; for electrical connection it is then sufficient, for example, to bring the front sides, metallized by flame-spraying, directly into electrical contact as shown in Fig. 3.2-5b. So that the height of such a stack of cylindrical windings does not become too large, one can also provide an internal series connection in addition to the external one. For this purpose several electrode foils are inserted into the winding one after another, as indicated in Fig. 3.2-6. Only the first and last electrode foil is led out as a terminal in each case.

3.2.2 Design of wound capacitors

a) Capacitance of a winding

The basic manufacturing process of a wound capacitor is shown in Fig. 3.2-7. The tapes comprising 4 layers for example, are wound one above the other on the mandrel. Let the dielectric have a layer thickness s and the overlapping width of both the metal foils be B. For the capacitance of the winding, besides s and ϵ_r, it is only B as well as the length L of the upper metal foil, called the measuring foil, that are significant. Double the effective capacitance is achieved by the winding process, so that for the winding capacitance we have:

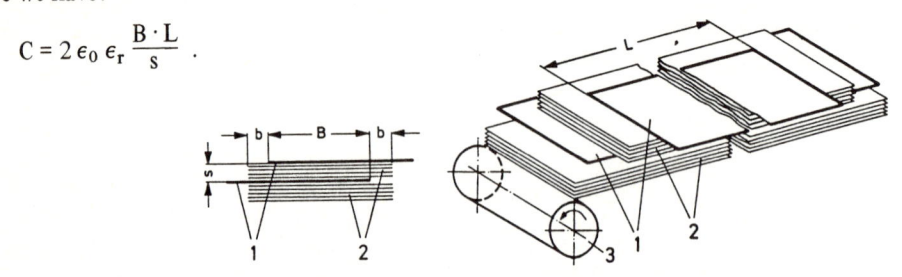

$$C = 2 \epsilon_0 \epsilon_r \frac{B \cdot L}{s} .$$

Fig. 3.2-7 Manufacture of a wound capacitor
1 metal foil, 2 dielectric, 3 mandrel

The edge spacing b is to prevent an external flashover and is usually chosen to be about 5...10 mm.

b) Dielectric

Special capacitor paper of 10...20 μm thickness combined with an impregnating medium, plastic films (polypropylene, polyethylene, styroflex), or also a mixture of both, is used as dielectric.

The individual layers cannot be made with an arbitrarily low weak spot probability; a pervading weak spot in the dielectric would, however, have a breakdown as consequence. To prevent this, numerous layers are always arranged one above the other. The probability that in each one of the n layers a weak spot occurs at the same site is then only W_1^n, where W_1 is the probability that a certain elemental area of a layer represents a weak spot. For plastic films one therefore chooses 2 layers if possible and for paper, on account of its higher number of weak spots, 3 to 6 layers.

For measuring and coupling capacitors, to obtain low temperature dependence of the capacitance, the use of mineral oil impregnated paper dielectric is preferred and one achieves $\epsilon_r \approx 4.2$ and $\tan \delta < 0.2\%$. With persisting overstepping of the inception voltage, X-wax formation and ultimate breakdown at the electrode edges should be expected.

For power capacitors, paper or paper/foil dielectric is used, hitherto chiefly impregnated with chlorinated diphenyl (askarel). The achievable power density, because $\epsilon_r \approx 5.5$ and due to the chosen slightly higher operating field strength, is significantly greater than that for oil as impregnating medium, with about the same dissipation factor. With askarel however, on account of the danger of hydrochloric acid formation at the metal edges,

3.2 High-voltage capacitors

special attention must be paid to ensure that the inception voltage is not exceeded, even for short durations if possible. Further, in place of askarel, other impregnating media (e.g. isopropylbiphenyl), better from the environmental control viewpoint, are increasingly being used, although they do have a substantially lower dielectric constant (ϵ_r = 2.7), yet permit higher operating field strengths.

In impulse capacitors, in the interest of a high energy density, one tolerates short-duration overshooting of the inception voltage. Compared with mineral oil, castor oil has also proved useful for the impregnation of the paper windings, when $\epsilon_r \approx 5$ can be obtained. The high dissipation factor of almost 1% does not interfere here.

In special cases even gases can be of technical significance as impregnating media, namely when plastic films are used. For example with polyethylene films one achieves an extraordinarily low-loss dielectric with $\tan \delta < 10^{-4}$ and $\epsilon_r \approx 1.5$ [*Brand, Kind* 1972; *Brand* 1973].

c) Operating field strengths of alternating voltage and direct voltage capacitors

The most important dimensional factor of a high-voltage capacitor is the operating field strength E because it appears quadratically in the expressions for power and energy density. For alternating voltage, besides the long-term ageing mechanisms, the inception voltage determines the upper limit. An approximate calculation is given in Section 1.3.4b. According to this, for a configuration in mineral oil at a pressure of 1 bar, we have:

$$U_e = 30 \left(\frac{s}{r}\right)^{0.5} \quad \text{in kV for s in cm}.$$

From this the inception voltage U_e for a winding with oil-impregnated paper dielectric, made of 5 layers each 10 μm thick and ϵ_r = 4.2, can be calculated to be $U_e \approx 1$ kV. The effective operating field strength must, therefore, lie safely below 20 V/μm. Experience shows that with askarel paper one can choose a slightly higher field strength than with oil-impregnated paper.

For direct voltage, edge effects have no real significance; particularly for the behaviour during direct voltage testing it is in general the breakdown voltage of the winding which is more critical. Fig. 3.2-8 shows, for different thicknesses of the individual layers of an askarel insulated paper dielectric, the breakdown field strength E_d as a function of s [*Guthmann* 1954]. As the measurements indicate, for a given thickness s of the di-

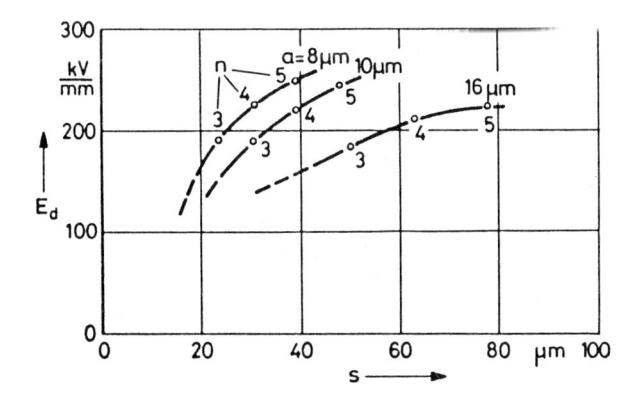

Fig. 3.2-8 Breakdown strength E_d of askarel impregnated paper under direct voltage

s = thickness of the dielectric
a = paper thickness
n = no. of layer

electric E_d increases strongly with the use of thinner paper. The increase in E_d as a consequence of lower weak-spot probability for a larger number of layers is also seen from the measured values of the average breakdown field strength $\overline{E}_d$ during direct voltage testing of askarel-impregnated paper windings [*Liebscher, Held* 1968]:

No. of layers	2	3	4
$\overline{E}_d$ in V/μm	100	170	230

The paper here is a type of density 1.2 g/cm^3 and thickness $10...60\,\mu\text{m}$.

The requirement that the dielectric shall be subdivided as often as possible naturally holds for all voltage stresses. With regard to their manufacture and processability, capacitor paper is available down to a thickness of only about $6\,\mu\text{m}$. One must therefore compromise between optimum utilization of the dielectric, by as thin a paper layer as possible, and the manufacturing costs which increase strongly with decreasing paper thickness.

For alternating voltage applications the following effective field strengths are chosen:

Power capacitors
(askarel-impregnated paper) $E = 15...20 \text{ V/}\mu\text{m}$
(askarel-impregnated paper/foils) $E = 35...40 \text{ V/}\mu\text{m}$

Measuring and coupling capacitors
(mineral oil impregnated paper) $E = 10...15 \text{ V/}\mu\text{m}$

For direct voltage application an example for the usual operating field strength is:

Smoothing capacitors
(mineral oil impregnated paper) $E = 80...100 \text{ V/}\mu\text{m}$

The range quoted in each case indicates that the acceptable values depend very strongly upon the construction used. The preparation of the dielectric in particular, discussed in detail in Section 2.5, as well as the careful manufacture and connection of the windings are very important. In new constructions it is always necessary to convince oneself, by means of comprehensive life-time testing, of the correctly chosen operating field strength.

d) Design of impulse capacitors

The sudden discharge of a high-voltage capacitor constitutes high mechanical stressing of the windings and their connections. This must be taken into account in impulse capacitors particularly, but also has practical significance for capacitors for direct and alternating voltages in the event of an external flashover.

The electrostatic surface force shown in Fig. 3.2-9a can be calculated using the energy equation:

$$F' = \frac{1}{2} \epsilon_0 \epsilon_r E^2 \ .$$

If it is assumed that $E = 75 \text{ V/}\mu\text{m}$ and $\epsilon_r = 4.2$, it follows that $F' \approx 10^5 \text{ N/m}^2 = 1$ bar. For a sudden discharge this pressure, withstood earlier in the winding, vanishes very abruptly; for an oscillating discharge pressure oscillations at twice the discharge frequency occur, and the consequence is a strong mechanical alternating stress on the thin metal and insulating foils.

3.2 High-voltage capacitors

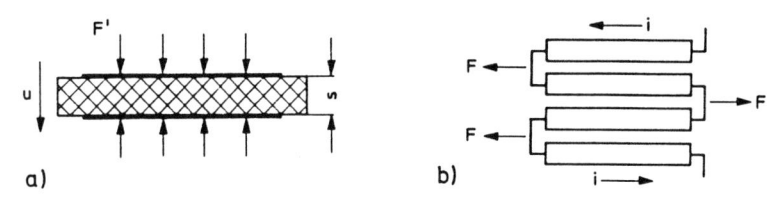

Fig. 3.2-9 Electro-mechanical forces in a wound capacitor
U, i instantaneous values of voltage and current, a) electrostatic surface force F',
b) electromagnetic force F on the winding connections

Fig. 3.2-10
Oscillatory discharge of a capacitor
U_0 = charging voltage
k = oscillatory factor

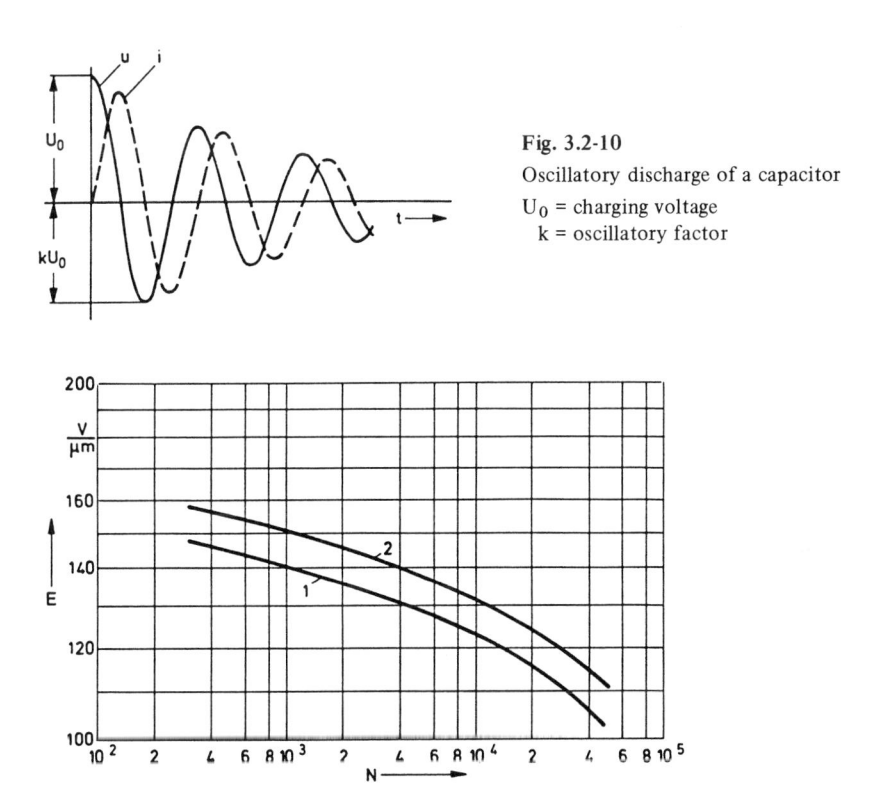

Fig. 3.2-11 Lifetime of impulse capacitors with oil-impregnated paper dielectric (k = 0.8) 1: s = 5 · 12 μm, 2: s = 4 · 12 μm

The electromagnetic forces indicated in Fig. 3.2-9b make it essential that the construction of the winding connections be of high mechanical quality and strength. To keep the unavoidable magnetic fields induced by the discharge current i at a minimum, the impulse capacitors must have direct axial current leads at each winding of the metal foils if possible. Impulse capacitors are subject to operational oscillatory discharges as shown in Fig. 3.2-10. For the stresses so produced the oscillatory factor k is of great significance. In impulse capacitors for plasma physics experiments, it is usual to aim for a lifetime of about N = 10,000 discharges for k = 0.8. Fig. 3.2-11 gives an idea of the relation between the operating field strength and the discharge number N to be expected up to breakdown [*Naglik* 1973]. The values were obtained for oil-impregnated paper windings and are valid for k = 0.8; for k = 0.5 one would expect about 4 times the number of discharges to breakdown. It can be seen that a decrease in the operating field strength by 10% also permits anticipation of a 4 times longer lifetime; for the choice of castor oil in place of mineral oil, a further increase in the lifetime may be expected. Usual values of operating field strength of impulse capacitors lie, as in direct voltage capacitors, at $E = 80...100 V/\mu m$.

3.2.3 Types of design

Power capacitors are constructed preferentially with flat windings, pressed into cubic packets and tightly enclosed in sheet metal housing. In this way good heat transfer is guaranteed and the temperature dependent expansion of the impregnating medium is taken up by the membrane effect of the housing walls. Since power capacitors are usually required for medium voltages and with two-pole insulation, the leads for the connecting terminals as well as the insulation of the windings against the housing can be managed well. Fig. 3.2-12 shows, as an example, a unit for 10 kV, 100 kVA.

Coupling capacitors, as well as the capacitor parts of capacitive voltage transformers, are required for voltages from 110 kV upwards and for comparatively low power ratings, so that heating up presents no problem. They are therefore mainly built in an insulated housing type construction with porcelain housing. Fig. 3.2-13 shows an example of a coupling capacitor for 110 kV, 4400 pF. In place of the flat windings shown, cylindrical windings with internal series connection are also chosen. The expansion of the dielectric which is sealed off from the surrounding air is taken up either by a gas cushion, as shown in the figure, or by bellows.

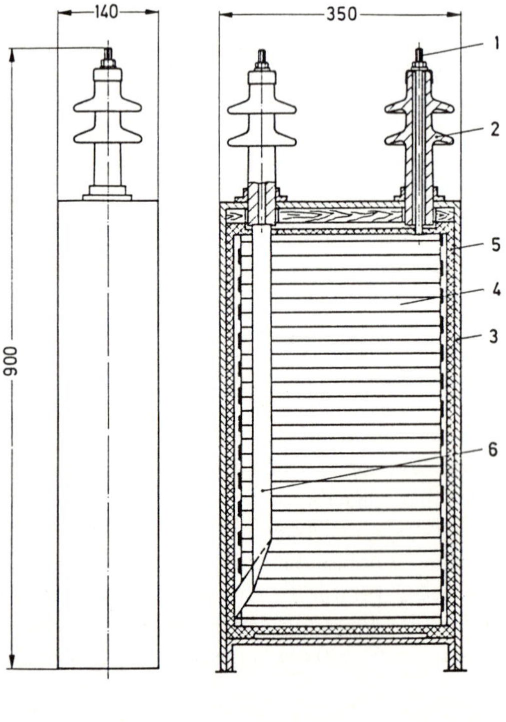

Fig. 3.2-12
50 Hz power capacitor for 10 kV, 100 kVA
1 connecting bolt
2 porcelain bushing insulator
3 steel housing
4 flat winding
5 paper or pressboard insulation
6 return lead

3.2 High-voltage capacitors

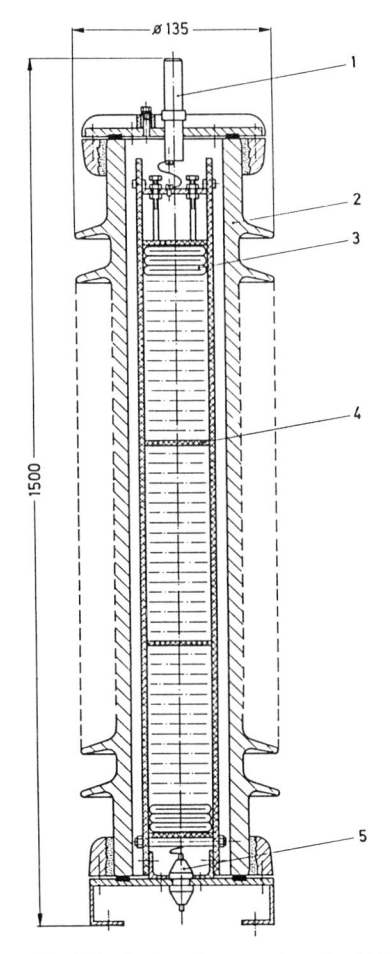

Fig. 3.2-13 Coupling capacitor for 110 kV, 4400 pF (MWB)

1 terminal bolt
2 porcelain housing
3 flat winding
4 pressing gibs
5 bushing insulator for low-voltage connection

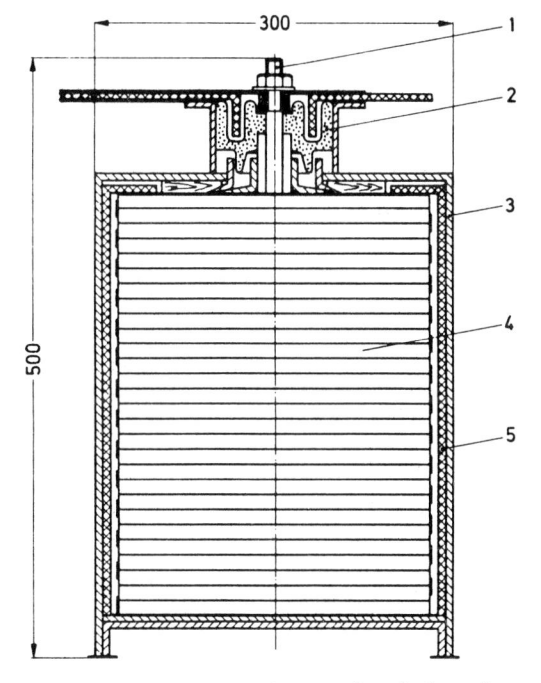

Fig. 3.2-14 Low inductance impulse capacitor 40 kV, 1.5 kWs (Siemens)

1 strip conductor terminal
2 bushing insulator
3 steel housing
4 flat winding
5 paper or pressboard insulation

Impulse capacitors must have low inductance in the arrangement of windings and at the terminals. Whilst in capacitors for impulse voltage generators in the voltage range over 200 kV constructions with an insulated housing can still be found, impulse capacitors for low inductance capacitive energy storage must be built with coaxial or strip terminals [*Liebscher, Held* 1968]. As an example Fig. 3.2-14 shows an impulse capacitor for 40 kV and 1.5 kWs. This differs from the power capacitor shown above in that it has a low inductance strip conductor terminal and current return via the housing.

3.3 Bushings and lead-outs[1])

The problem to be solved in bushings and lead-outs, namely the insulated penetration of one electrode by another has already been mentioned in Section 3.1.1 in connection with insulator types (Fig. 3.1-7c and d). As a rule, an electrically safe high voltage carrying conductor is to be passed through as small an opening as possible in an earthed plane and a rigid mechanical connection is also very often required. Appropriate to their technical significance, only designs for alternating and impulse voltages shall be discussed here. For direct voltage the possibility of capacitive voltage control, well proven for time-variant voltage, is naturally not available, but the risk of gliding discharges is appreciably lower.

3.3.1 Basic configurations

a) Bushings

The mutual interaction between insulating material and construction already emphasised becomes particularly clear in the case of bushings. The same problem leads to completely different solutions for different materials. Bushings for operating alternating voltages up to about 30 kV are made of porcelain or cast resin; for high voltages, insulating bodies made of hardboard or soft paper in the wound construction and with porcelain housing are preferred.

Fig. 3.3-1 shows the principle of gliding discharge suppression at the flange for a porcelain bushing. The porcelain is provided with a conductive coating under the flange by a metal spraying technique; the coating terminates in a protruding edge for field control purposes. To prevent partial discharges in any air spaces between the shaft and the insulating material, either sufficient clearance must be maintained between the shaft and porcelain or, as shown in the figure, a conductive coating must be applied on the inside too.

In an insulating body made of cast resin as in Fig. 3.3-2, the shaft or its surrounding metal tube, through which the current leads pass, is directly embedded; partial discharges on the

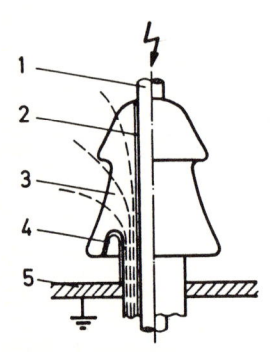

Fig. 3.3-1 Field control in a porcelain bushing
1 shaft, 2 inner conductive coating, 3 insulating body, 4 outer conductive coating, 5 flange

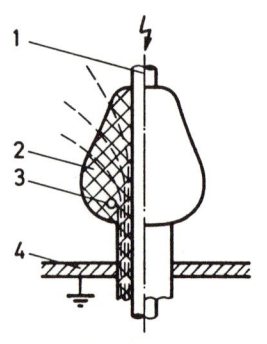

Fig. 3.3-2 Field control in a cast resin bushing
1 shaft, 2 insulating body, 3 extended electrode, 4 flange

---

[1]) Comprehensive treatment in [*Böning* 1955; *Sirotinski* 1958; *Philippow* 1966].

3.3 Bushings and lead-outs

high-voltage conductor are thus prevented. The gliding discharge problem is solved by the embedded protruding electrode. Then, with the proper choice of insulator profile, the field strength at the inclined boundary layer to the air is decreased to permissible values. With this arrangement, in contrast to the porcelain configuration, no pre-discharges occur prior to flashover when the voltage is increased. On account of the easy embedding of metals, cast resin offers diverse possibilities to the constructing engineer.

Bushings for rated voltages over 60 kV are usually built with capacitive voltage grading by intermediate electrodes, since otherwise a very large flange diameter would have to be chosen. The principle of this kind of capacitive grading was expressed by *R. Nagel* in 1906 for transformer bushings. Here the potentials at the boundary layer between the insulating material and the surrounding medium are capacitively graded by intermediate electrodes (metal foils) inside the insulating material. Fig. 3.3-3 shows the effect of grading foils on the potential distribution of a cylindrical bushing, where for reasons of clarity the radial dimensions are disproportionately large. The diameter and length of the intermediate electrodes must be designed so that the desired potentials on the metal foils are produced by way of the partial capacitances. Calculation procedures for this are given in Section 3.3.2.

The principle of capacitive grading is at first glance independent of the type of insulating material, but here too, the relation between the inception voltage at a sharp electrode edge and the thickness of the insulating material underneath it, shown in Fig. 3.2-2, must be noted. If one aims to avoid cumbersome screening electrodes for field control at the edges between the electrodes, a small insulation thickness must be chosen.

This leads to a finely graduated capacitive grading which can practically only be realized in insulating materials manufactured from thin bands by the winding technique. Apart from certain exceptions, capacitive gradings are therefore only made of hardboard or soft paper, and plastic films.

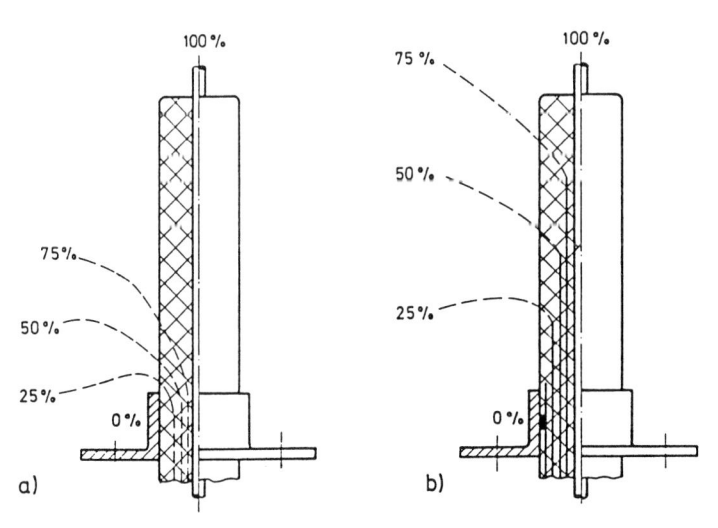

Fig. 3.3-3 Potential distribution in cylindrical bushings
a) without intermediate electrodes, b) with intermediate electrodes (capacitor bushing)

b) Lead-outs

In contrast to bushings, lead-outs by themselves cannot be thought of as a constructional element since they are inseparably connected to a high-voltage apparatus. An important example for this is the cable termination [*Lücking* 1981], which has the special difficulty that it can be subjected to a high-voltage test only in conjunction with the cable on site.

Nowadays, for the medium voltage range, plastic cables are predominantly used. The principle of a lead-out suitable for this purpose is shown in Fig. 3.3-4. In the region of the termination the cable housing is removed and either a prefabricated conical insulating piece is inserted over the conductor insulation or attached on site by winding plastic tapes or sometimes, even by casting. By shaping the edge of the earthed foil appropriately, one can achieve useful uniformity of the field strength at the insulator surface.

For operating voltages over 110 kV, cable terminations are chiefly produced with a finely graduated capacitive grading, whose basic construction is demonstrated in Fig. 3.3-5. The termination is detached from the metal housing and a conical insulating piece is either inserted over it or made on site. This contains a finely graduated capacitive grading made by winding insulating tapes with metal foils interposed. The use of this kind of potential grading allows the cable termination to have a relatively small external diameter, even for very high voltages.

As an example for an equipment lead-out, the potential control of an instrument transformer coil for high voltage, insulated with oil-impregnated paper, is shown in Fig. 3.3-6. In the region of the core at earth potential, the high-voltage coil must have an earthed conductive layer on the outer surface. So that no gliding discharges can occur at its edge, this must be covered either with an earthed and bandaged control ring or the coil must be given a finely graduated capacitive grading (see also Fig. 3.5-8).

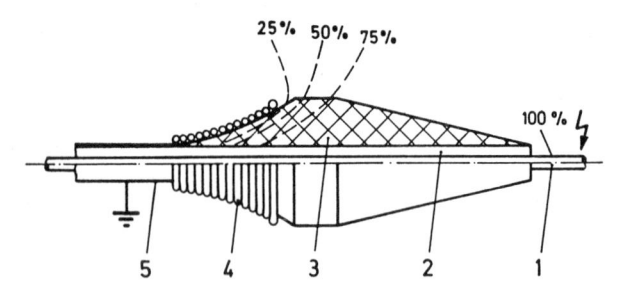

Fig. 3.3-4 Principle of a cable lead-out without capacitive grading

1 conductor
2 cable without housing
3 insulating body
4 wire bandage with edge electrode
5 cable housing

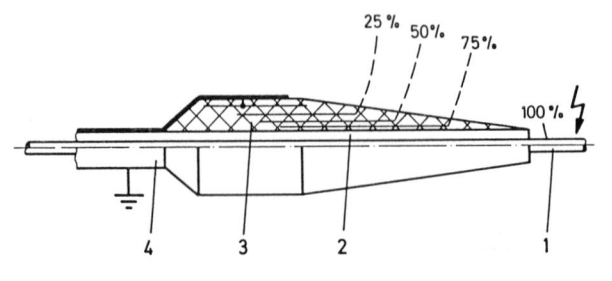

Fig. 3.3-5 Principle of a cable lead-out with capacitive grading

1 conductor
2 cable without housing
3 insulator with capacitive grading
4 cable housing

3.3 Bushings and lead-outs 123

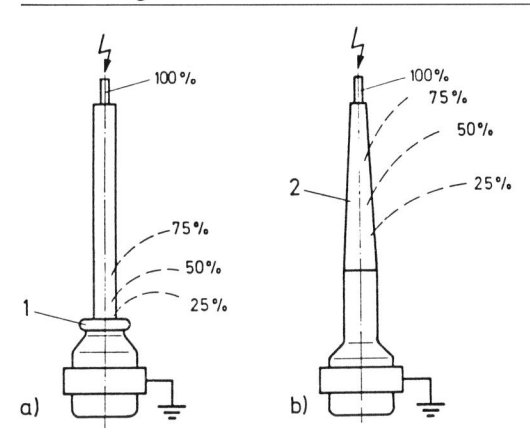

Fig. 3.3-6 Lead-out of an instrument transformer coil
a) with control ring 1
b) with capacitive grading 2

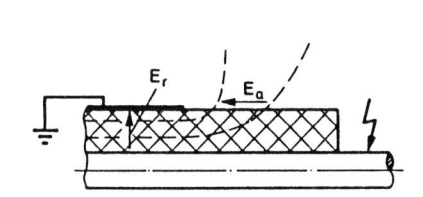

Fig. 3.3-7 Electric stress in bushings and lead-outs
E_r = radial field strength
E_a = axial field strength

3.3.2 Calculation of capacitive gradings

The insulation of a bushing or lead-out is stressed, as shown in Fig. 3.3-7, radially and axially, where above all the boundary surface between the insulating material and the surrounding medium should be considered a critical area. The radial component E_r of the electric field strength can cause breakdown of the insulating material, whilst the axial component E_a can under certain circumstances lead to surface discharges along the boundary surface. Since the electric strength of the insulating material stressed to breakdown limit is appreciably higher than that of the boundary layer stressed to flashover limit, the axial stress is in general far more critical.

a) Calculation of inception voltage

In capacitive gradings, as a rule, the spacing of the layers is very much smaller than the corresponding diameter; so one can apply the plane model of the electrode placed on the surface, as in Section 1.3.4, to calculate the inception voltage. The a. c. inception voltage for partial discharges at the edges of the conductive layers is calculated as in the case of wound capacitors approximately to be

$$U_e = K \left(\frac{s}{\epsilon_r}\right)^{0.5} \quad \text{in kV for s in cm}.$$

Here the following guiding values can be assumed for K [*Böning* 1955; *Brand* 1973]:

Configuration		K
metal edge	in air	8
	in SF$_6$	21
metal or graphite edge	in oil	30
graphite edge	in air	12

One should note that for the choice of resistive layers (graphite paper) with graphite edge the inception voltage for alternating voltage can be appreciably increased, but the grading of the impulse voltage may then possibly become impaired due to the large currents which then flow in the axial direction [*Kappeler* 1968]. In every case the onset of high-energy brush discharges should be calculated with K = 80 approximately, these should not even occur temporarily, e.g. during high-voltage testing, since then the insulating material would become permanently damaged. For optimum utilization of the dielectric it is recommended that the capacitive gradings be arranged so that the same partial voltage

$$\Delta U = \frac{U}{N}$$

is across two adjacent layers, where N is the total number of layers. This condition is presumed in the following.

b) Double-sided capacitive gradings

Double-sided capacitive gradings are needed for bushings. The symbols in Fig. 3.3-8 shall be taken for the calculation. The layers are numbered from the central bolt (n = 0) to the flange (n = N). Let the spacing between the layers be

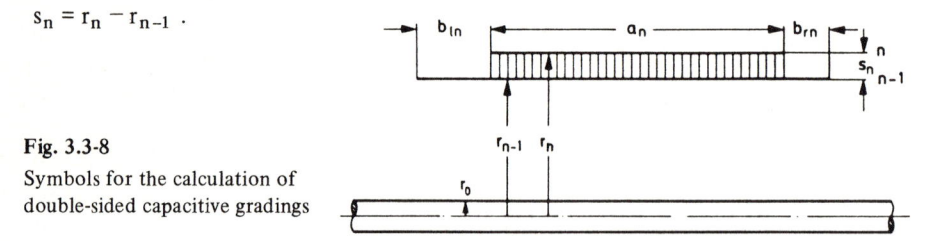

$$s_n = r_n - r_{n-1} .$$

Fig. 3.3-8
Symbols for the calculation of double-sided capacitive gradings

If the edge overlap on the left side b_{ln} is equal to that on the right side b_{rn}, a symmetrical bushing is obtained. For different surrounding dielectrics however, bushings are required with differently graded lengths on either side.

The main capacitance represented by the vertically hatched portion can be calculated, neglecting the edge fields, to be:

$$C_n = \frac{2\pi \epsilon_0 \epsilon_r a_n}{\ln r_n/r_{n-1}} .$$

Since all the capacitors C_n are connected in series, these must be equal to one another because ΔU = const:

$$C_n = \text{const.} = C .$$

Now one must decide whether the potential should be graded in the radial or axial direction.

For radial grading, the radial field strength $E_r = \Delta U/s_n$ must be constant. This is satisfied for a constant layer spacing s_n. Assuming $C_{n+1} = C_n$, the recursion formula follows:

$$a_{n+1} = a_n \frac{\ln r_{n+1}/r_n}{\ln r_n/r_{n-1}} .$$

3.3 Bushings and lead-outs

For the plane approximation, with $s_n = r_n - r_{n-1} \ll r_n$, we have:

$$a_{n+1} \approx a_n \frac{r_{n-1}}{r_n}.$$

These recursion formulae permit the determination of the next layer from the data of the preceding one. The innermost radius R_0 is usually given; from the maximum permissible radial field strength E_r, the spacing s_n is obtained for given voltage. If a_0 is also given, then all further data can be calculated. The resulting envelope of the layer edges is hyperbolic in shape, as shown in Fig. 3.3-9 for an asymmetrical bushing.

For axial grading the axial field strength $E_a = \Delta U/b_n$ must be constant. It then follows that the layer ends on each side must be displaced by the same length:

$$b_{ln} = \text{const} = b_l, \quad b_{rn} = \text{const} = b_r.$$

For the length of the layer we then have:

$$a_{n+1} = a_n - b_l - b_r.$$

From the assumption $C_{n+1} = C_n$ it follows:

$$\ln \frac{r_{n+1}}{r_n} = \frac{a_{n+1}}{a_n} \ln \frac{r_n}{r_{n-1}}.$$

The plane approximation yields:

$$s_{n+1} \approx s_n \frac{a_{n+1}}{a_n} \frac{r_n}{r_{n-1}}.$$

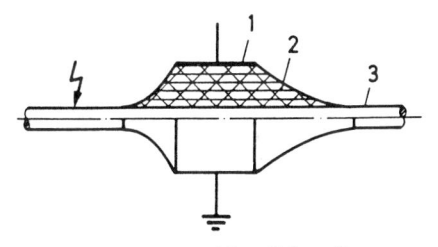

Fig. 3.3-9 Bushing with radial grading
1 flange, 2 layers, 3 central bolt

The dimensions of the succeeding layers once again result from those of the preceding ones. An example for the contour of the layers of this kind of bushing is shown in Fig. 3.3-10. The flashover length L indicated is calculated approximately as:

$$L \approx N b_r.$$

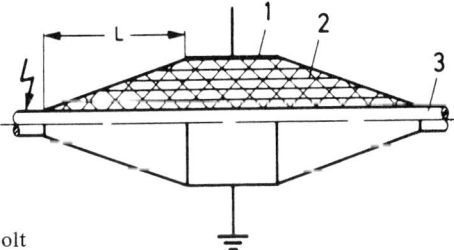

Fig. 3.3-10
Bushing with axial grading
1 flange, 2 layers, 3 central bolt

The large number of steps involved make a calculation scheme advisable. To calculate a bushing with axial grading the procedure is as follows:

1. Choice of the number of layers N based upon the experience that the voltage across two layers during a.c. test voltage U_p shall be about 12 kV.
   For example, the test voltage for 110 kV bushings is $U_p = 260$ kV; from this it follows that the number of layers N = 22.

2. Choice of the flashover length L from the condition that at U_p, the average field strength along the boundary surface must be limited. In air, one can assume about

3...4 kV/cm; under oil, depending upon the construction of the oil component, two to four times higher field strengths are permissible [*Kappeler* 1968]. This and the previous condition give the edge overlaps b_r and b_l.

3. As a rule, the bolt radius r_0 and the total length a_0 are given. One then chooses an initial value r_1 and continues to calculate with the recursion formula.
4. Finally, one must also make a few control calculations. On the one hand, the highest radial field strength E_r for the test voltage may not exceed the strength of the insulating material and, on the other hand, the highest permissible operating voltage must be distinctly less than U_e.

c) One-sided capacitive grading

For the calculation of one-sided gradings, the symbols drawn in the scheme of Fig. 3.3-11a shall be adopted. In addition to the main capacitance C_n (hatched region between adjacent layers), stray capacitances C_{sn} to the bolt also appear. With regard to the practical significance, only axial grading with identical lengths for all the steps viz. $b_n = b$ and $a_{sn} = a_s$ shall be treated.

At the junction n in Fig. 3.3.-11b the current equation is:

$$\Delta U \omega C_{n+1} = \Delta U \omega C_n + n \Delta U \omega C_{sn} .$$

It follows that

$$C_{n+1} = C_n + n C_{sn}$$

and then, applying the formula for a cylindrical capacitor, we have:

$$\frac{a_n - b}{\ln r_{n+1}/r_n} = \frac{a_n - a_s}{\ln r_n/r_{n-1}} + \frac{n a_s}{\ln r_n/r_0} .$$

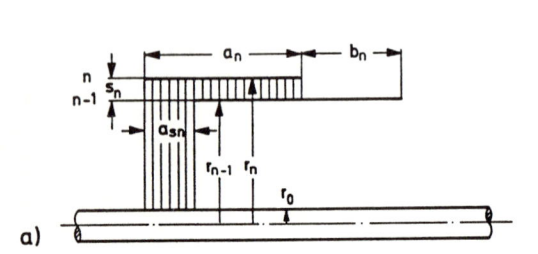

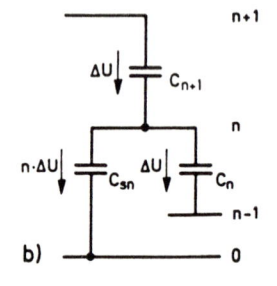

Fig. 3.3-11 Calculation of one-sided capacitance grading
a) symbols, b) equivalent circuit

If one takes into account that $a_{n+1} = a_n - b + a_s$, then

$$\ln \frac{r_{n+1}}{r_n} = \frac{a_n - b}{\dfrac{a_n - a_s}{\ln r_n/r_{n-1}} + n \dfrac{a_s}{\ln r_n/r_0}}$$

can be used as a recursion formula.

The calculation is analogous to that for the two-sided axially graded bushing. First one determines N and b. Then a_s and a_1 are chosen. With given r_0 and chosen r_1 the calculation can begin.

3.3 Bushings and lead-outs 127

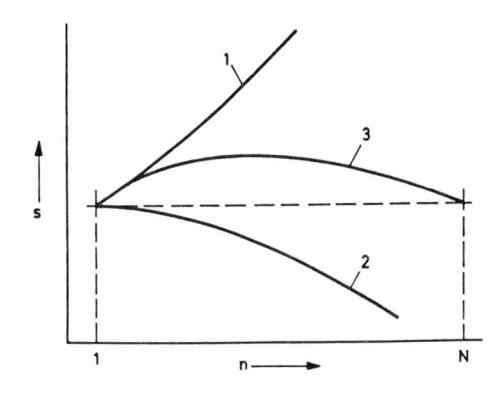

Fig. 3.3-12
Control diagram for calculations of one-sided gradings

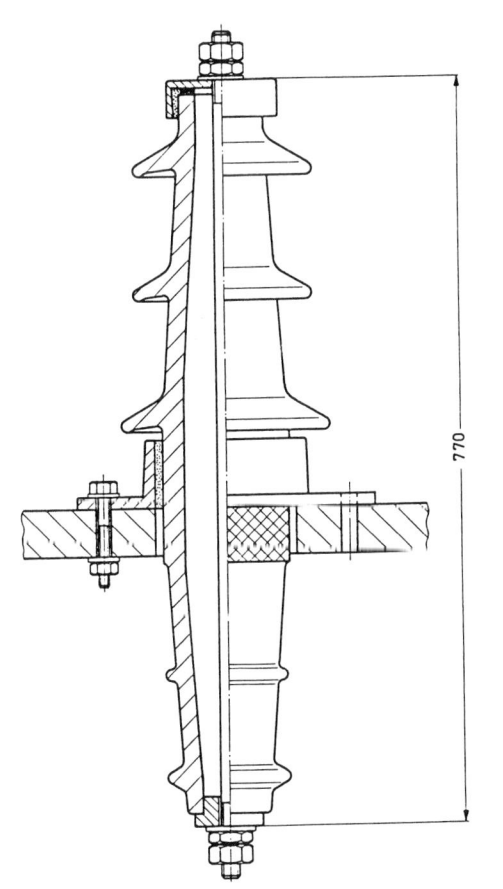

Fig. 3.3-13 20 kV porcelain bushing for outdoor/indoor application (Rosenthal/Driescher)

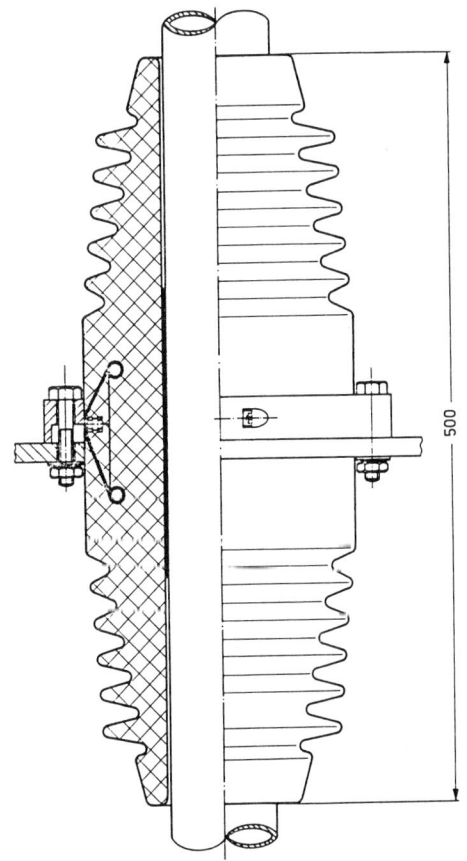

Fig. 3.3-14 Indoor bushing for medium voltage (20 kV) with epoxy resin (Calor–EMAG)

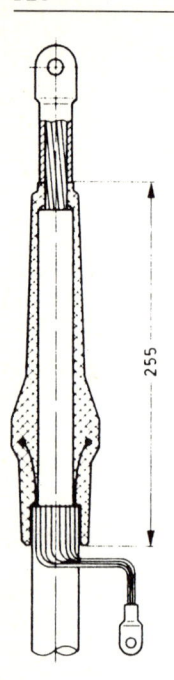

Fig. 3.3-15 Termination for 20 kV PE cable (Siemens)

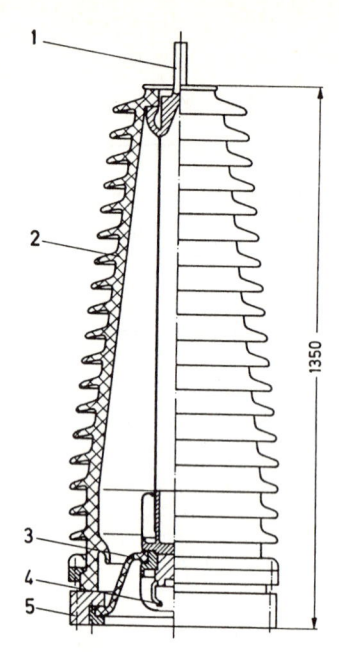

Fig. 3.3-16 SF_6-outdoor bushing for 110 kV station (BBC)
1 connecting bolt
2 porcelain housing
3 support insulator
4 isolator terminal
5 tubular flange

To help recognize badly chosen initial values, one can draw the diagram represented in Fig. 3.3-12. The layer spacings s_n are plotted against the number of layers n. If the values increase too rapidly (1), the bushing is too thick and the material poorly utilized. For decreasing values (2), the radial field strength becomes too high. Curve (3) shows the best behaviour since initial and final values s_1 and s_N are approximately equal. For high-voltage gradings, to save time, it is advisable to set up a computer programme.

3.3.3 Types of design

In the medium voltage range bushings are mostly required for switching stations. The insulating material is porcelain; in the majority of indoor applications however, epoxy resin is primarily chosen, which enables much smaller dimensions. Fig. 3.3-13 shows a 20 kV porcelain bushing of the outdoor or indoor type. The conductor at high voltage is introduced later and care must be taken to see that discharges in the air space around the bolt are prevented during operation. The 30 kV epoxy resin indoor bushing reproduced in Fig. 3.3-14 comprises a cast grading electrode in the region of the flange and a metal tube through which the conductor can be introduced.

The termination for a 20 kV PE cable as shown in Fig. 3.3-15 also solves the problem of field grading by the embedded electrode; here, a pre-fabricated insulating piece of epoxy

3.3 Bushings and lead-outs

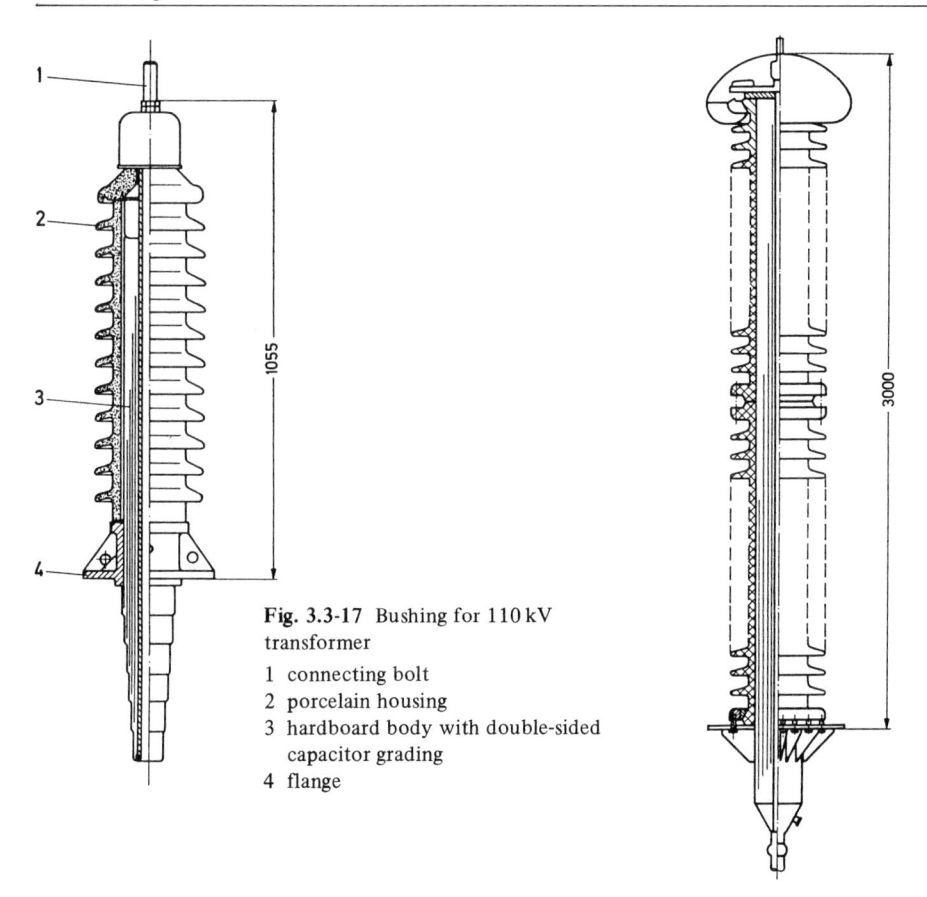

Fig. 3.3-17 Bushing for 110 kV transformer

1 connecting bolt
2 porcelain housing
3 hardboard body with double-sided capacitor grading
4 flange

Fig. 3.3-18 Termination for 400 kV oil-filled cable

Fig. 3.3-19 Cable lead-in for a 400 kV transformer (Micafil)

resin with a funnel-shaped grading electrode is inserted over the cable termination after removal of the cable housing. Then the inner conductor is connected to the cable terminal and the grading ring to the earthing terminal. Sometimes these lead-outs for plastic cables are also made by winding on thermoplastic insulating bands on site.

In gas-insulated switching stations bushings even for voltages of 110kV can still be designed without capacitive grading. Fig. 3.3-16 shows that by appropriate shaping of the electrodes, a simple lead-out for SF_6-insulated substations can be realised.

In general, however, in the voltage range over 110kV, one can no longer do without finely graduated capacitor grading; the construction engineer is then obliged to use an insulating body manufactured by the winding process. A typical example is the 110kV transformer lead-out as in Fig. 3.3-17. A hardboard roll with capacitor foils is surrounded by outdoor porcelain housing and the clearing gap filled with an insulating compound which prevents discharges in the housing interior. The 400kV termination for an oil-filled cable shown in Fig. 3.3-18 is designed similarly. On the cable termination freed of its cable housing, a soft paper roll containing the capacitor grading is inserted or wound on site. The entire arrangement is then impregnated and filled with oil.

Finally, Fig. 3.3-19 shows how much space can be saved when a high-voltage cable is connected directly to a transformer. The avoidance of air gaps in the transformer bushing as well as in the cable termination is of great importance, particularly for indoor installations, as in the case of unit transformers in underground power houses.

3.4 Transformer windings

For the dimensions of power transformers the magnetic, electric and thermal fields are important. The transformer actually works by the combined action of the windings and the iron core. Particularly in the region of the windings strong electric fields develop in transformers, where high temperatures and short-circuit forces can also act simultaneously. The insulation system must therefore not only guarantee sufficient electric strength but also be able to take up the mechanical forces and dissipate the loss heat.

In this section a few special insulation problems of transformer windings shall be discussed, where, besides plastic insulation systems, above all oil-impregnated paper insulation systems deserve our interest in view of their great practical significance. As a prerequisite for the design of transformers a few properties of the magnetic circuit and windings shall be discussed first.

3.4.1 Design factors for magnetic circuits

In this section some basic relationships for the design of alternating current circuits with magnetic windings will be given. Here, only the fundamental component of the time-variant quantities shall be considered in each case[1]).

a) The laws of induction and magnetomotive force

The simplest form of a winding with a closed magnetic core is shown in Fig. 3.4-1. An alternating voltage $\underline{U}$ applied to a winding of w turns results in a magnetizing current $\underline{I}_0$.

---

[1]) Comprehensive treatment in [*Küchler* 1966; *Philippow* 1966; *Richter* 1967; *Taegen* 1970].

3.4 Transformer windings

This current excites a magnetic flux Φ in the core with a constant effective iron cross-section q_E and a mean iron path l_E; flux leakage shall not occur. The magnetic induction or flux density $\underline{B}$ in the core for uniform distribution over q_E is:

$$\underline{B} = \frac{\underline{\Phi}}{q_E}.$$

Fig. 3.4-1
Winding with a closed magnetic core

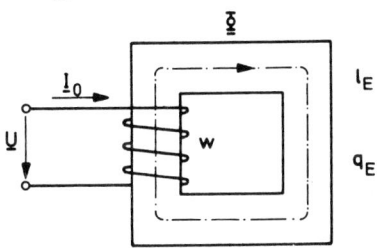

For sinusoids with angular frequency $\omega = 2\pi f$, it follows from the law of induction, in complex notation,

$$\underline{U} = w\, q_E\, j\, \omega \underline{B}.$$

Normally, the peak value is quoted for the magnetic induction. So for the modulus we have:

$$\hat{B} = \sqrt{2}\, B = \sqrt{2}\, \frac{U}{w\, \omega\, q_E}.$$

For $f = 50\,\text{Hz}$, U in V and q_E in m^2, we have:

$$\hat{B} = \frac{U/w}{222\, q_E} \quad \text{in T}.$$

U/w is the voltage between turns. As a rule the iron core of transformers consists of grain oriented magnetic sheets of 0.35 mm thickness made of silicon-iron alloys. The lamination causes q_E to be somewhat smaller than the geometrical iron core cross-section; the space factor is about 0.96.

The permissible rated flux density $\hat{B}_n$ in equipment has an upper limit due to the iron losses, the inrush current, as well as the magnitude and distortion of the magnetizing current. The following values are chosen:

in power transformers $\qquad \hat{B}_n = 1.5 \ldots 1.8\,\text{T}$,
in testing transformers $\qquad \hat{B}_n = 1.2 \ldots 1.5\,\text{T}$,
in inductive voltage transformers $\hat{B}_n = 0.7 \ldots 1.2\,\text{T}$.

In current transformers the flux density depends on the current to be measured and the load; further, the flux density at rated current and rated load depends on the proposed application of the transformer (see Section 3.5). Here we have:

$$\hat{B}_n = 0.03 \ldots 0.3\,\text{T}.$$

The law of magnetomotive force relates the magnetic field strength $\underline{H}$ to the magnetizing current $\underline{I}_0$. For sinusoids we have:

$$\underline{I}_0\, w = \underline{H}\, l_E.$$

It follows:

$$\underline{I}_0 = \frac{\underline{H}\, l_E}{w}.$$

The relation between $\underline{H}$ and $\underline{B}$ is the magnetization characteristic of the core material. For the moduli both the representations

$$\hat{B} = f(H)$$

or

$$\hat{B} = \mu_0 \mu_r H \sqrt{2}$$

with

$$\mu_r = \mu_r(\hat{B}),$$

are used; μ_0 is the permeability of vacuum.

Using these quantities, the modulus of the magnetizing current is calculated as follows:

$$I_0 = B \frac{l_E}{\mu_0 \mu_r w} = \frac{U \, l_E}{\omega w^2 \, q_E \, \mu_0 \mu_r}.$$

The magnetization characteristic of common ferromagnetic core materials is highly non-linear. For sinusoidal induction therefore, an extremely distorted magnetizing current results. In addition $\hat{B}$ and $\underline{H}$ are not in phase on account of the magnetizing losses. This can be accounted for by the formulation[1])

$$\underline{H} = H_w - j H_b$$

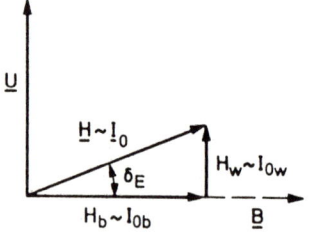

Fig. 3.4-2
Phasor diagram of electric and magnetic quantities

where H_w is the component of $\underline{H}$ in the direction of $\underline{U}$ (Fig. 3.4-2); the same representation is also valid for the magnetizing current $\underline{I}_0 = I_{0w} - j I_{0b}$. The angle δ_E shown in the figure is designated the iron loss angle and corresponds to the dielectric loss angle of an insulation. We have

$$I_{0w} = I_0 \sin \delta_E, \quad I_{0b} = I_0 \cos \delta_E.$$

The magnetizing losses P_{Fe} of the core are obtained from the electrical quantities

$$P_{Fe} = U I_{0w} = \omega B H_w l_E q_E.$$

For the specific losses referred to unit mass of density γ it follows:

$$P'_{Fe} = \frac{1}{\gamma} \omega B H_w.$$

For silicon-iron alloys one can take $\gamma = 7600 \, \text{kg/m}^3$, for nickel-iron alloys $\gamma = 8100...8700$ kg/m³. In the manufacturers' data $\hat{B} = f(H)$ and $P'_{Fe} = f(\hat{B})$ are usually given; the real component of the magnetizing current can then be calculated as:

$$I_{0w} = \frac{H_w l_E}{w} = \sqrt{2} \, \frac{P'_{Fe} l_E}{\omega B w} \gamma.$$

---

[1]) The separation of components representation shown here is preferred to the other possibilities for accounting for iron losses, such as the statement $\mu_r = \mu' - j \mu''$.

3.4 Transformer windings

The curves $\hat{B} = f(H)$, $P'_{Fe} = f(\hat{B})$, $\delta_E = f(\hat{B})$ and $\mu_r = f(\hat{B})$ are reproduced in Appendix 4 for a few important core materials. One typical example for a grain oriented silicon-iron alloy and for a nickel-iron alloy respectively was chosen. Practical designs must naturally be based on more exact values for each of the materials used [e.g. *Vakuumschmelze* 1977].

b) The short-circuit impedance of windings

Transformers possess at least two windings through both of which the main flux passes. Fig. 3.4-3 shows the basic construction of a transformer with two coaxial cylindrical windings (1, 2) and a stepped laminated core (3), as well as the corresponding circuit diagram.

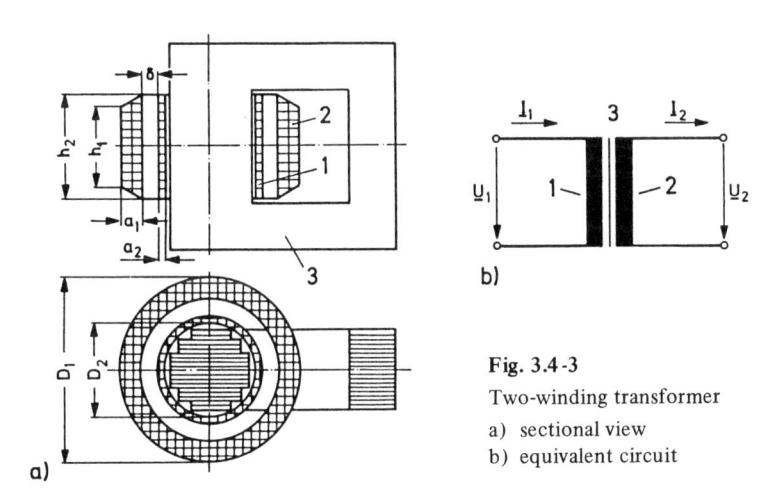

Fig. 3.4-3
Two-winding transformer
a) sectional view
b) equivalent circuit

The 4-pole equivalent circuit of this simple two-winding arrangement contains lumped effective resistances and inductances. It can be represented, as in Fig. 3.4-4a, using an ideal transformer to account for the transformation given by the number of turns w_1 and w_2, or by referring all the elements to one of the windings as in Fig. 3.4-4b. Z_0 follows from the components of the magnetizing current[1]) calculated in a).

For the components of the impedances of the equivalent circuit attributed to both windings the following relationships are valid:

$$\underline{Z}_1 = R_1 + j X_{S1}$$
$$\underline{Z}_2 = R_2 + j X_{S2} .$$

The magnitudes of R_1 and R_2 are obtained from the effective resistances of the windings, taking the additional losses due to current displacement into consideration (skin effect). In solid conductors of large cross-section these can assume impermissibly high values, for which reason twisted conductors are often used in such cases.

---

[1]) The T-equivalent circuit used here is strictly only valid in the case that the entire leakage flux can be distributed unequivocally between both windings; this is not so in technical transformers. Nevertheless, consistent with the literature [e.g. *Schlosser* 1963; *Richter* 1963], it shall be employed for the simple considerations of this book.

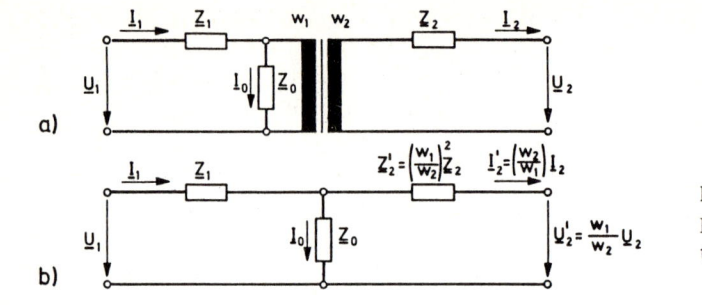

Fig. 3.4-4 Equivalent circuits of a two-winding transformer

The values of the leakage reactances X_{S1} and X_{S2} are derived from the leakage flux. For the calculations, not reproduced here, one assumes that at the front sides of the core, at the level of the yoke, magnetically infinitely good conducting planes exist [*Taegen* 1970]. The flux lines of the leakage field produced during short-circuit are then parallel to the core axis. Now part of the leakage flux can be attributed to each winding, where the distribution is effected arbitrarily at half the width of the leakage canal between windings. Using the symbols of Fig. 3.4-3a, we have, approximately;

$$X_{S1} = \frac{\mu_0 \omega \pi D_m}{h_m} \left(\frac{\delta}{2} + \frac{a_1}{3} \right) w_1^2$$

$$X_{S2} = \frac{\mu_0 \omega \pi D_m}{h_m} \left(\frac{\delta}{2} + \frac{a_2}{3} \right) w_2^2 .$$

The abbreviations used here are:

$$h_m = \frac{1}{2}(h_1 + h_2)$$

and

$$D_m = \frac{1}{2}(D_1 + D_2) .$$

The total short-circuit reactance referred to the primary side is:

$$X_S = X_{S1} + X'_{S2} = \frac{\mu_0 \omega \pi D_m}{h_m} \left(\delta + \frac{a_1 + a_2}{3} \right) w_1^2 .$$

Finally, for the relative short-circuit voltage it follows with rated values of voltage U_n, current I_n and conductor losses P_{Cun}:

reactive part $\quad u_x = \dfrac{X_S I_n}{U_n} .$

real part $\quad u_r = \dfrac{(R_1 + R'_2) I_n}{U_n} = \dfrac{P_{Cun}}{U_n I_n} ,$

apparent value $\quad u_k = \sqrt{u_x^2 + u_r^2} .$

3.4 Transformer windings

Due to the assumptions made in the derivation, the above relationships for the leakage reactances are only approximations; the resulting deviations rarely lie above 10%. More exact methods for the calculation of the short-circuit reactances even of complicated arrangements are given in the literature [*Richter* 1963; *Philippow* 1966].

The circuit of Fig. 3.4-4a can be extended to multi-winding transformers, even when the windings do not have a common potential. In this manner the short-circuit reactances of testing transformers and inductive potential transformers in cascade connection, for instance, can be calculated from the values of the individual stages [*Kind* 1972]. For more exact investigations of multi-winding transformers, it is necessary to devise other equivalent circuits [*Schlosser* 1963].

3.4.2 Assembly and connection of windings

The windings of a transformer must be magnetically closely coupled to one another. If windings around the same core are arranged axially they are called disc windings; if they lie radially, cylindrical windings (Fig. 3.4-5). In each case, because of the electromagnetic forces on short-circuiting, the winding arrangement must be symmetrical.

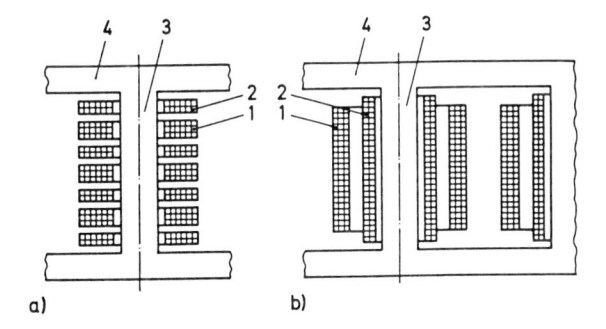

Fig. 3.4-5
Assembly of transformer windings
a) disc winding
b) cylindrical winding
1 high-voltage winding
2 low-voltage winding
3 core
4 yoke

For about the same conductor cross-section, the high-voltage winding requires larger insulating clearances to the yokes and also between individual turns. The lower stacking factor so caused is indicated in Fig. 3.4-5 by different winding cross-sections. In high-voltage transformers, for reasons of insulation expense, in general only cylindrical windings are used and these shall be discussed in the following.

In fact, high-voltage windings can be built up of individual coils or individual layers, each connected in series. In the first case the coils are arranged axially and in the second case the layers are arranged radially.

The most important assemblies of coil windings are shown in Fig. 3.4-6. In the single-coil connection, completely identically wound coils are connected in series, where passing the connecting lead through the space between the coils raises special insulation problems; between the two coils the full voltage of each single coil lies at points with the same radius. In the preferred double-coil connection the problem of the connecting lead does not arise, but the voltage between two points on adjacent coils with the same radius varies from zero to double the coil voltage. In contrast to the single-coil connection, the individual coils of a double-coil connection cannot be wound continuously but must be wired subsequently. An important special case of coil windings is the (inverted)

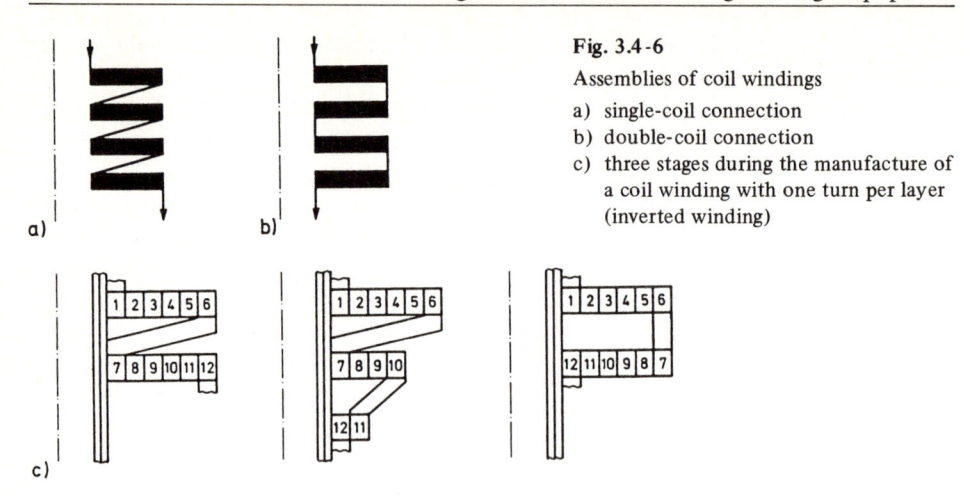

Fig. 3.4-6

Assemblies of coil windings
a) single-coil connection
b) double-coil connection
c) three stages during the manufacture of a coil winding with one turn per layer (inverted winding)

winding shown in Fig. 3.4-6c. It is used when the conductor cross-section permits every coil itself to consist of individual windings wound directly over each other. The three phases of manufacture shown are: continuous winding of two coils; rearrangement ("inversion") of the second coil, for which purpose its turns on the winding former must be loosened; reconstruction of the second coil in reversed conductor sequence. In this manner one obtains coil windings in double-coil connection without soldered connections.

The most important assemblies of layer windings are shown in Fig. 3.4-7. Once again one must differentiate between a series connection along the shortest path, corresponding to the double-layer connection shown, and a return lead to the opposite side. It is not expedient to pass the connecting leads between the layers as in the single-layer connection with inner connections, but rather on the outside with outer connections.

It should once more be noted that for the double-layer scheme, twice the layer voltage appears between two layers at the sides lying opposite the connection.

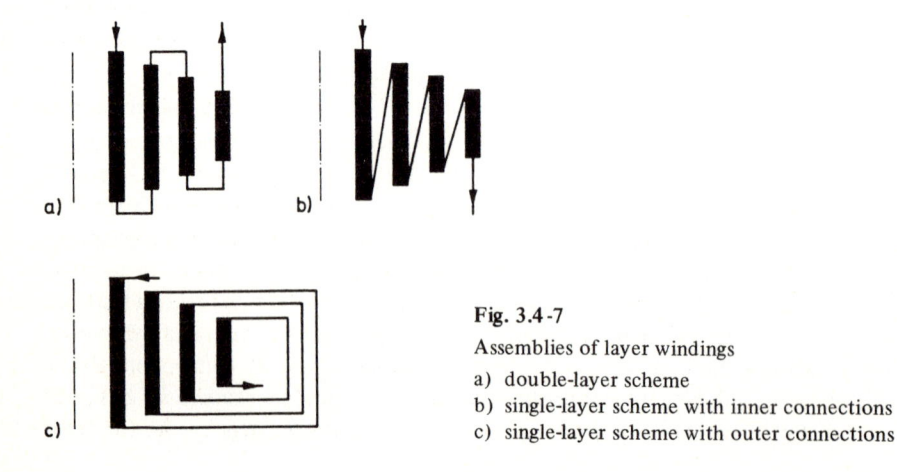

Fig. 3.4-7

Assemblies of layer windings
a) double-layer scheme
b) single-layer scheme with inner connections
c) single-layer scheme with outer connections

3.4 Transformer windings

3.4.3 Insulation of high-voltage windings

In the insulation of cylindrical windings against the core and other windings, essentially three characteristic insulation problems occur for which certain denotations are usual:

high-voltage winding against low-voltage winding:

    major or leakage canal insulation,

high-voltage winding against yoke:

    yoke or end insulation,

high-voltage winding against high-voltage winding for three-phase construction:

    window insulation.

These three areas are designated winding insulation because here the high-voltage winding as a whole appears as an electrode to be insulated. The purpose of the layer insulation or coil insulation is to insulate the individual layers or coils against one another. Finally, one must also remember the insulation of individual turns, described as the inter-turn insulation.

Power transformers are predominantly insulated with oil-impregnated paper, the good properties of which, especially for large power ratings and high voltages, are not achieved by any other insulation. For medium voltage and indoor applications dry transformers with epoxy resin insulation are often used as well, since oil transformers in the event of damage are a great fire risk. In the following, exclusively oil transformers shall be discussed.

To insulate the turns, for small cross-sections up to a few mm^2, varnished copper wires with circular or rectangular cross-section are used. For larger cross-sections a paper insulated rectangular conductor is always used. For large currents stranded conductors made of varnished partial conductors with a common paper insulation are also used, to prevent additional losses. As paper insulation for the individual conductors, 0.02...0.2mm thick paper tapes for example are suitable, wound in semi-overlap.

To insulate the layers or coils, one chooses pressboard or soft paper with oil ducts for heat disposal. During a.c. test voltage stresses of about 50...100 kV/cm in the oil ducts are permissible.

For the winding insulation, laminated insulations made of pressboard or soft paper with oil ducts are common. Dimensioning is such that for a temporary breakdown of the oil duct the remaining dielectric can still withstand.

Hardboard is avoided for high voltages where possible since pressboard (transformer board) possesses high mechanical strengths too and at the same time very good electric strength.

An example for the winding insulation of a transformer is shown in Fig. 3.4-8. Between the high-voltage and low-voltage windings 1 and 2 lies the leakage canal which is subdivided by pressboard tubes 3 so that no continuous free oil duct develops. A circular shielding electrode 4 is placed on each of the windings to make the electric field at the front end of the windings uniform. These shielding rings are at the potential of the last turn and must either be split or made of resistance material so that they do not constitute a short-circuited winding. They are bandaged with soft paper. The winding and pressboard tubes are secured mechanically by barrier-like angle rings 5 made of press-

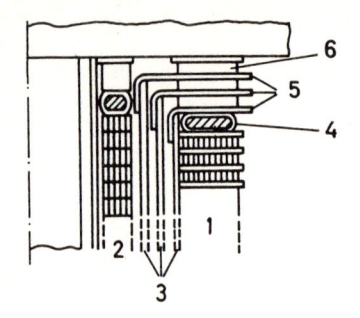

Fig. 3.4-8
Example for winding insulation of an oil transformer
1 high-voltage winding
2 low-voltage winding
3 pressboard tubes
4 shielding rings
5 angle-rings
6 spacer blocks

board and spacer blocks 6. Since oil circulation must be guaranteed, these are arranged at the appropriate spacing.

3.4.4 Impulse voltage performance and winding construction

a) Estimation of the initial voltage distribution

High-voltage windings of transformers are intricate structures, being weakly damped networks comprising capacitances, self and mutual inductances. If an impulse voltage appears across the terminals of the transformer, characteristic oscillations are excited which can result in impermissibly high local stresses of the insulation inside the windings. For this reason, extensive investigations have been carried out to determine the voltage distributions occurring in different types of winding [e.g. *Müller* 1975 and the literature quoted therein].

A greatly simplified, but still useful assumption is that in the first instant of stress with a steep impulse voltage only the capacitances of the network are effective. A simple equivalent circuit for the calculation of the initial distribution is shown in Fig. 3.4-9. It is a homogeneous iterative network containing only the main capacitances C as well as the earth capacitances C_e of a high-voltage winding [*Strigel* 1955]. The voltage distribution along the winding after applying a step voltage U_0 is calculated for the earthed neutral point ($U_n = 0$) as follows:

$$U_\nu = U_0 \frac{\sinh(n-\nu)\alpha}{\sinh n\alpha} \ .$$

For the free neutral point ($I_n = 0$) we have:

$$U_\nu = U_0 \frac{\cosh(n-\nu)\alpha}{\cosh n\alpha} \ .$$

In these expressions $\alpha = \sqrt{C_e/C}$.

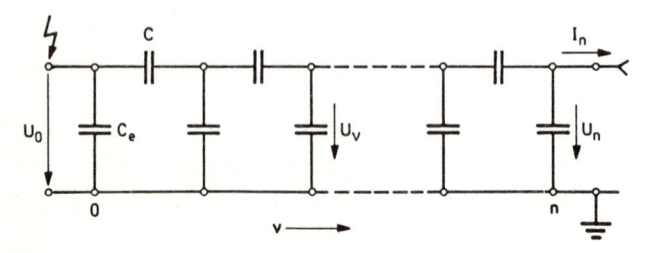

Fig. 3.4-9
Capacitive equivalent circuit to calculate the initial distribution

3.4 Transformer windings

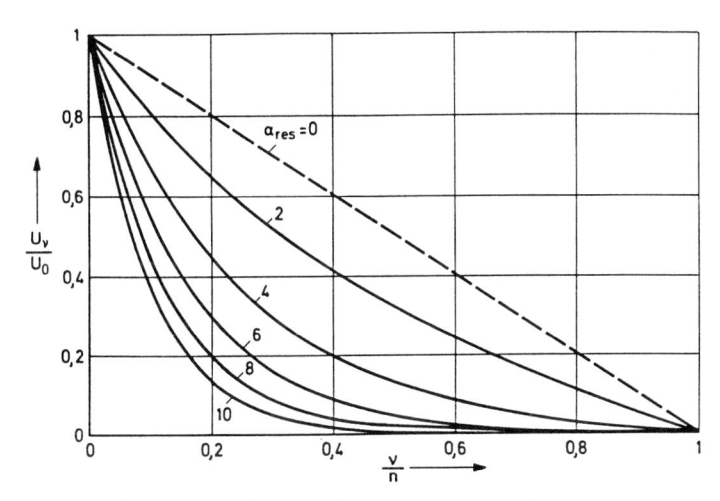

Fig. 3.4-10 Initial distribution with earthed neutral point

To demonstrate, let us introduce the resultant earth capacitance $C_{e\,res} = nC_e$ and the resultant main capacitance $C_{res} = C/n$. We then have:

$$\alpha_{res} = n\alpha = \sqrt{\frac{C_{e\,res}}{C_{res}}}.$$

Theoretical initial distributions for various values of α_{res} for earthed neutral point are shown in Fig. 3.4-10. (Usual values are $\alpha_{res} = 2...10$.)

If the maximum stress on a section of the winding is calculated, we have, approximately, ($\sinh \alpha_{res} \approx \cosh \alpha_{res}$) for the earthed neutral point and for the free neutral point:

$$\left(\frac{dU_\nu}{d\nu}\right)_{max} = \left(\frac{dU_\nu}{d\nu}\right)_{\nu=0} = -\frac{U_0}{n}\alpha_{res}.$$

Here, α_{res} is immediately evident as a factor of the overstress on the first winding section compared with the ideal linear voltage distribution which reappears as the normal operating condition after a transient mechanism has faded out. When designing a transformer winding it is therefore necessary to aim for as small a ratio of the earth capacitance to the main capacitance as possible.

The deviation of the initial distribution from the final distribution is of great significance to the impulse voltage stressing of a winding during the subsequent transient mechanism. Since damping of the oscillations is not possible because of the low losses required during normal operation, one must aim not to excite dangerous oscillations at all. The less the initial and final distributions differ from one another, the smaller the possibility of oscillations occurring. One speaks in such a case of a low-oscillation transformer.

b) Measures to improve the impulse voltage performance

Estimation has shown that the input turns of the winding are overloaded. One would therefore initially think of reinforcing the insulation of the input turns. This measure should be regarded with caution however, since the capacitances between the relevant turns would then be simultaneously reduced and the component voltage would increase. A measure which is advantageous in every case is to augment the coupling capacitances of the input turns to the high-voltage electrode with the aid of the shielding rings shown in Fig. 3.4-8, or with additionally arranged insulated shields connected to high-voltage potential. The most effective measure, however, is augmentation of the main capacitance by using a shielded layer winding or, for coil windings, by so-called interleaved winding. Moreover, in layer windings one can achieve an additional increase of the main capacitance by a doubly concentric construction of the winding, with one low-voltage winding inside and one outside, as shown in Fig. 3.4-11. A type of construction usual in power transformers for high voltages are interleaved coil windings, whose schematic arrangement is shown in Fig. 3.4-12. Besides the series connection of each run schown, parallel connections are also possible. Naturally, the frequent crossover which becomes necessary causes additional construction work and this considerably affects the manufacturing costs of transformers.

A normal inverted winding, as in Fig. 3.4-6c, has $\alpha_{res} = 12$; with interleaved windings, as in Fig. 3.4-12, one obtaines values of $\alpha_{res} = 4$ and in special designs up to $\alpha_{res} = 2$ [*Brechna* 1958].

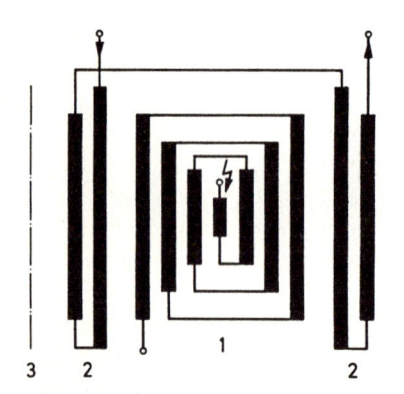

Fig. 3.4-11 Scheme of a doubly-concentric layer winding
1 high-voltage winding
2 low-voltage winding
3 winding axis

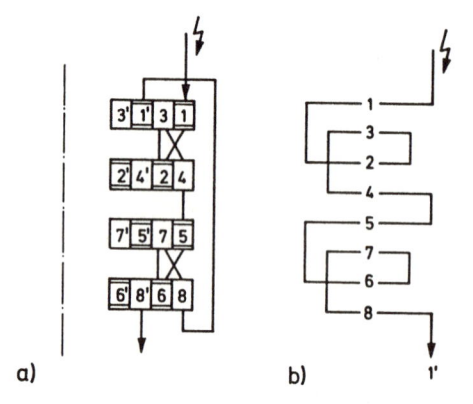

Fig. 3.4-12 a) Scheme of an interleaved winding with 2 runs (the numbering corresponds to the successive windings)
b) Current paths in the high-voltage windings

3.4.5 Types of transformer winding

For the insulation of transformer windings several designs are chosen depending on the degree of stress to which the windings are subject. The design details will be described here using the example of two transformers whose insulation is of oil with barriers and oil-impregnated paper. In the first example we have chosen a coil winding and in the second a layer winding.

3.4 Transformer windings

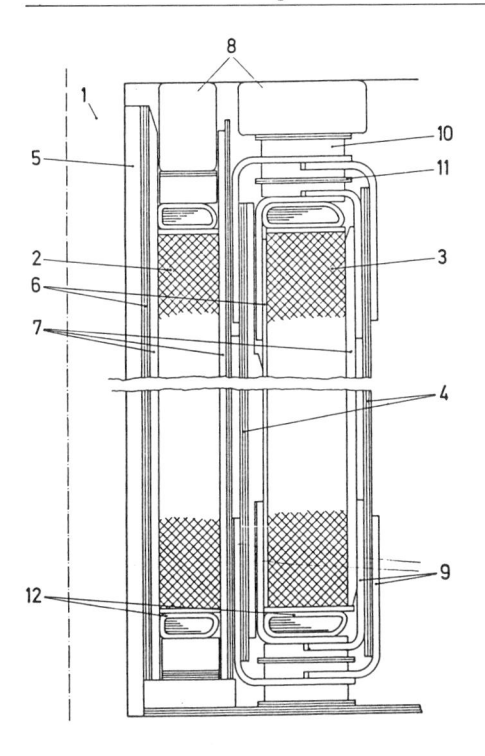

Fig. 3.4-13

Transformer with angle-ring insulation and coil winding

1. iron core
2. low-voltage winding
3. high-voltage winding
4. pressboard cylinder
5. wooden rods
6. support cylinder
7. spacer bars
8. pressure rings
9. angle-rings
10. spacer blocks
11. pressboard discs
12. shielding rings

Fig. 3.4-13 shows a transformer section with angle-ring insulation and coil winding, taken through the iron core 1, the windings 2 and 3 of a phase and a window. The main insulation is of oil and oil-impregnated pressboard cylinders 4 which surround the low-voltage and high-voltage windings and have an angular profile to ensure oil entry at the front end. The support cylinder 6 of the low-voltage winding 2 is propped radially against the core 1 by the wooden rods 5. The low-voltage winding itself is supported radially by spacer bars 7 against the support cylinder. These spacer bars, which also lie between the low-voltage and high-voltage windings, must possess cooling oil ducts between the neighbouring construction elements and be in a position to sustain the electromagnetic forces acting between the windings during a short-circuit. Under short-circuit the windings try to enlarge the leakage canal between them. The clamping force which the pressing rings 8 exert on the windings should be about as large as the contracting force of the windings in the axial direction acting during short-circuit, so that the winding composite cannot loosen itself. The windings are further supported by pressure rings 9, spacer blocks 10 and pressboard discs 11, the diameter of which must be larger than that of the spacer blocks, since otherwise short creepage paths could be caused. The pressure rings, usually made of pressboard, approximately follow the equipotential lines of the electric field. The shielding rings 12 effect a reduction of the field at the front ends of the winding.

Instead of pressure rings for the insulation of the windings one may also choose a soft paper insulation system, the so-called splayed flange. Fig. 3.4-14 shows a transformer section with splayed flange insulation and layer winding. The low-voltage winding 2 consists of two layers and the high-voltage winding 3 consists of four layers. In the layer

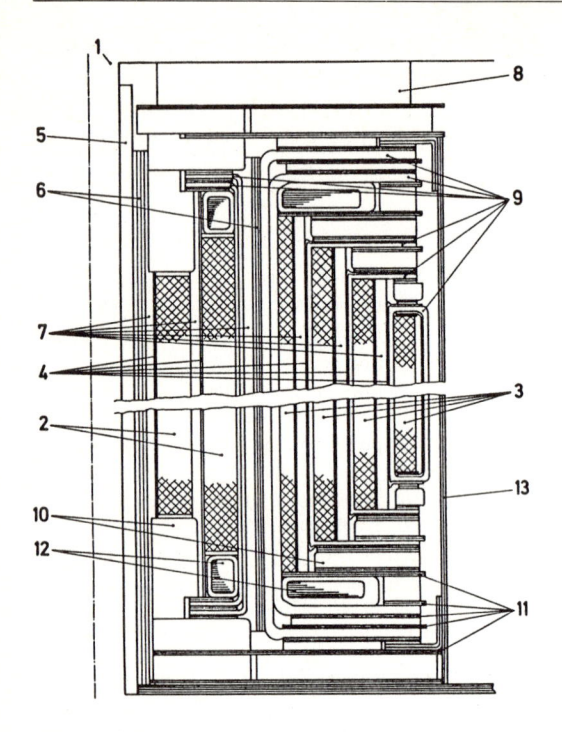

Fig. 3.4-14

Transformer with splayed flange insulating

1 iron core
2 low-voltage winding
3 high-voltage winding
4 pressboard cylinder
5 wooden rods
6 support cylinder
7 spacer bars
8 pressure rings
9 splayed flange insulation
10 spacer blocks
11 pressboard discs
12 shielding rings

windings assumed here more sophisticated internal insulation is necessary than in a comparable transformer with coil winding. During the manufacture of a splayed flange the ends of the wound paper cylinders are cut axially and arranged around the ends of the windings. Thus this insulation can be given an optimum fit to match the shape of the equipotential surfaces.

3.5 Instrument transformers[1])

The tasks of an instrument transformer are to transform currents and voltages to be measured into conveniently measurable secondary quantities, and to isolate the potential between the primary circuit and the secondary measuring circuit.

In voltage transformers the usual secondary rated voltage is 100 V or $100/\sqrt{3}$ V, whereas in current transformers very often a secondary rated current of 5 A or 1 A is specified. This standardization allows equipment for connection to such transformers (indicating or recording devices, energy meters, protective equipment) to be designed in a similar manner, independent of the value of the primary quantity.

Instrument transformers can be classified depending on their tasks as transformers for measuring purposes (operational and commercial measurements) and as transformers for protective purposes. The highest accuracy is demanded for commercial measurements:

---

[1]) Summary in [*Bauer* 1953; *Hermstein* 1969; *Zinn* 1977]. The most important properties of instrument transformers are specified in [VDE 0414 Part 1 to 5; IEC Publ. 185, 186, 44−3 and 44−4].

3.5 Instrument transformers

such transformers are made in the classes 0.1, 0.2 and 0.5. According to the verification laws they are, as a rule, subject to a verification test in electrical testing centres. More frequently these transformers are required to supply protective circuits. Here particularly in current transformers very high demands are made on their working range, since they must carry a multiple of the rated current depending upon the short-circuit current at the installation site.

The desire to be independent of external energy sources and the demands for high measuring performance makes the application of unconventional transformers, e.g. optoelectronic transformers, very difficult. Despite considerable efforts, see e.g. [*Müller* 1972], these transformers are therefore employed in only isolated cases: it is for this reason that only those types of instrument transformer generally used today are discussed here.

3.5.1 Inductive voltage transformers

The inductive voltage transformer functions like a lightly loaded transformer. The working range lies between 0.8 and 1.2 times the rated voltage as a rule. Transformers for protective purposes must however still comply with certain error limits, even for 0.05 and 1.9 times the rated voltage.

a) Circuitry

The circuit symbol and the circuit diagram of voltage transformers in a 3-phase network are shown in Fig. 3.5-1. One differentiates between single-pole and two-pole voltage transformers. The terminal denotation is standardized by VDE, as shown. Single-pole voltage transformers are simpler to construct than two-pole transformers because of the use of stepped insulation, since in the latter the entire high-voltage winding must be insulated against the earthed housing for the full test voltage. Two-pole transformers are therefore only used for rated voltages up to 30 kV, whereas single-pole transformers today are built for the highest voltages.

Single-pole voltage transformers in 3-phase networks, besides the measuring winding, have as a rule another winding for earth fault recognition, as shown in Fig. 3.5-2. During normal operation there is no voltage across the terminals of the series connection of the earth fault windings ($\Sigma U = 0$). However, if an earth fault occurs on one of the phases, a voltage ΣU three times the value of the phase voltage U_λ is generated. This voltage serves to excite or trigger the earth fault relays. The rated voltage of the earth fault winding is usually chosen so that for an earth fault ΣU reaches the same value as the line-to-line secondary voltage.

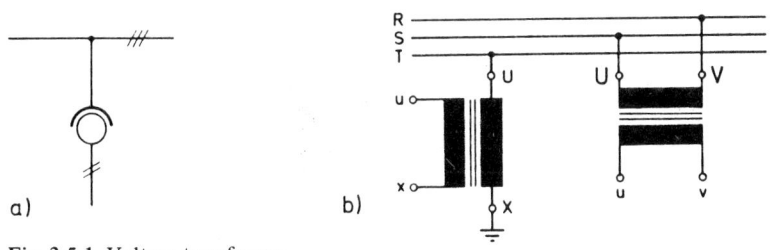

Fig. 3.5-1 Voltage transformer
a) circuit symbol, b) 1-pole and 2-pole designs

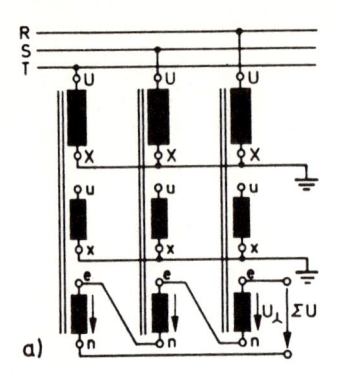

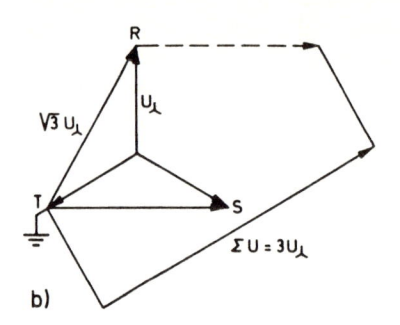

Fig. 3.5-2 Voltage transformer with earth fault winding in 3 phase network
a) circuit diagram, b) phasor diagram

b) Error calculation

The calculations for inductive voltage transformers are done on the basis of the transformer equivalent circuit (Fig. 3.5-3) discussed in Section 3.4.1; the quantities are transposed to the secondary side. The load impedance (burden impedance) $\underline{Z}_b$ comprises the impedance of the measuring system and that of the corresponding leads;

$$\underline{Z}_b = Z_b \, (\cos\beta + j \sin\beta)$$
$$= R_b + j X_b \, .$$

The error $\underline{F}$ is defined here as a complex quantity:

$$\underline{F} = \frac{\underline{U}_2' - \underline{U}_1}{\underline{U}_1} = \frac{\underline{U}_2 - \underline{U}_1'}{\underline{U}_1'} \, .$$

Going round the loop once we have:

$$\underline{U}_2 - \underline{U}_1' = -\underline{I}_0' \, \underline{Z}_1' - \underline{I}_2 \, (\underline{Z}_1' + \underline{Z}_2) \, ,$$

and so

$$\underline{F} = -\frac{\underline{I}_0'}{\underline{U}_1'} \underline{Z}_1' - \frac{\underline{I}_2}{\underline{U}_1'} (\underline{Z}_1' + \underline{Z}_2) \, .$$

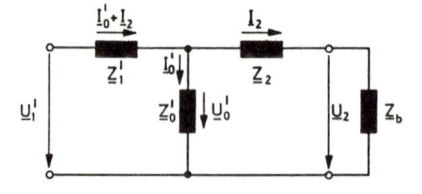

Fig. 3.5-3 Equivalent circuit of inductive voltage transformers

From the relationship derived in Section 3.4.1 for the magnetizing current it follows, with $U_0 \approx U_1$:

$$\underline{Z}_0 = \frac{U_1}{I_0} = \frac{\omega \, w_1^2 \, q_E \, \mu_0 \, \mu_r}{l_E} \, .$$

Since the error is very small in instrument transformers, one can neglect errors of the second order and substitute:

$$\underline{I}_2 \, \underline{Z}_b \approx \underline{U}_1' \, .$$

The error equation then simplifies to:

$$\underline{F} = -\frac{\underline{Z}_1}{\underline{Z}_0} - \frac{\underline{Z}_1' + \underline{Z}_2}{\underline{Z}_b} = \underline{F}_0 \, (U_1) + \underline{F}_b \, (R_b, X_b) \, .$$

3.5 Instrument transformers

The open-circuit error $\underline{F}_0$ contains the magnetizing current and is coupled to the voltage $\underline{U}_1$ via the magnetization characteristic. The loading error $\underline{F}_b$ depends only on the burden and thus on the loading of the transformer. To represent the phasor $\underline{F}$ in the complex plane it is expedient to choose the direction of $\underline{U}_1$ as the real axis ($\underline{U}_1 = U_1$). The real component so obtained is designated the voltage error F_u and the imaginary component the phase angle error $F_{\delta u}$:

$$\underline{F} = F_u + j F_{\delta u} .$$

The aim of the error calculation of an instrument transformer is now to determine F_u and $F_{\delta u}$ from the material properties with the geometrical parameters for various operating conditions (U_1, R_b, X_b). With the iron loss angle δ_E, one obtains as in Section 3.4.1

$$\underline{I}_0 = I_0 (\sin \delta_E - j \cos \delta_E)$$

and from this:

$$\underline{Z}_0 = Z_0 (\sin \delta_E + j \cos \delta_E) .$$

As an abbreviation let us introduce:

$$\underline{Z}_1 = R_1 + j X_{S1}$$
$$\underline{Z}'_1 + \underline{Z}_2 = R + j X_S .$$

The complete error equations in algebraic form are now;

$$F_u = -\frac{R_1 \sin \delta_E + X_{S1} \cos \delta_E}{Z_0} - \frac{R \cos \beta + X_S \sin \beta}{Z_b}$$

$$F_{\delta u} = -\frac{X_{S1} \sin \delta_E - R_1 \cos \delta_E}{Z_0} - \frac{X_S \cos \beta - R \sin \beta}{Z_b}$$

The phasor diagram of Fig. 3.5-4 shows the position of the most important voltages and currents with respect to one another as well as the error triangle indicated by hatching.

The end point of the phasor of the voltage $\underline{U}_2$, and with that the error $\underline{F}$, is usually described in polar coordinates by F_u in % and the error angle δu in minutes. Since an angle of $360° = 21,600'$ corresponds in radians to the value 2π, we have:

$$2\pi : 0.01 = 21,600' : 34.4'$$
$$1 \% \triangleq 34.4' .$$

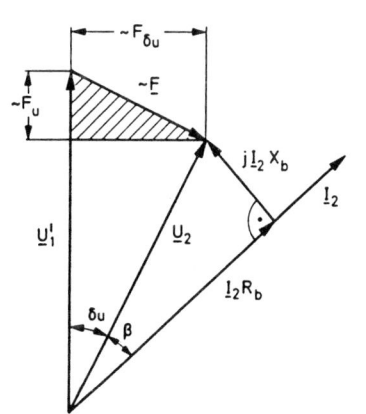

Fig. 3.5-4
Phasor diagram of the terminal voltages of an inductive voltage transformer

For the specified accuracies the errors are very small. To make the errors visible the phasor diagram should be very large. In the voltage transformer diagram according to J. A. Möllinger, shown in Fig. 3.5-5, the voltages $\underline{U}'_1$ and $\underline{U}_2$ are therefore drawn parallel and, moreover, all the phasors are referred to $\underline{U}'_1$. The origin of the coordinate system and the real axis coincide with the end of the phasor $\underline{U}'_1$. The voltage error F_u is positive if $U_2 > U'_1$; the phase angle error is positive if $\underline{U}_2$ leads the voltage $\underline{U}'_1$; hence positive values of $F_{\delta u}$ should be plotted to the left.

The open-circuit error $\underline{F}_0$ depends only on the magnitude of the voltage U_1. The loading error $\underline{F}_b$ is proportional to the secondary current I_2. Variation of the burden angle β for constant secondary current causes a rotation of the phasor $\underline{F}_b$ around the open-circuit point, so that the error follows the circular locus shown.

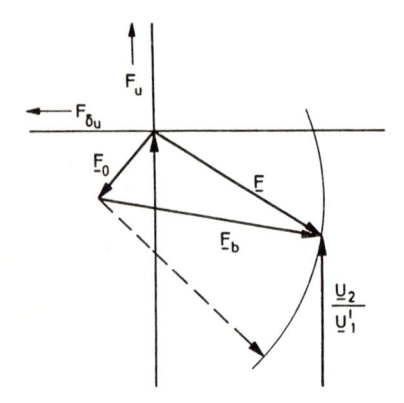

Fig. 3.5-5 Error representation in the voltage transformer diagram

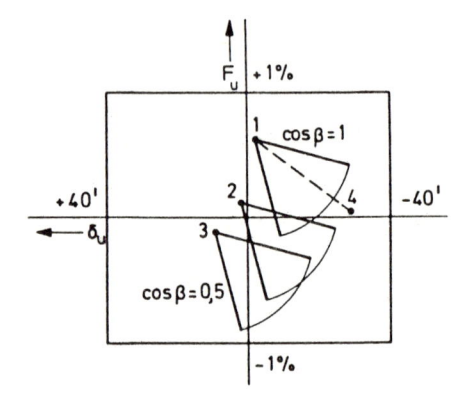

Fig. 3.5-6 Voltage transformer diagram with error limits according to class 1

In this kind of diagram one can indicate the error limits, as done in Fig. 3.5-6 for a voltage transformer of class 1, according to VDE 0414. By simple geometrical operations it is possible to determine exactly the effect of variation of the burden. If, for example, in the voltage range of 0.8 to 1.2 times the rated voltage, class 1 shall be maintained with $\cos \beta$ between 0.5 and 1.0, it is sufficient to specify the three open-circuit errors (points 1, 2, 3), and a further error (point 4) for any arbitrary known burden, to be able to draw the whole diagram.

The burden is usually not quoted as an impedance but as a reactive power which the impedance $\underline{Z}_b$ consumes at rated voltage on the secondary side. Let a burden of $P_{2n} = 120\,\text{VA}$ in a single-pole transformer with a secondary rated voltage $U_{2n} = 100/\sqrt{3}\,\text{V}$ serve as an example. The impedance is then:

$$Z_b = \frac{U_{2n}^2}{P_{2n}} = 28\,\Omega.$$

3.5 Instrument transformers

c) Designs

The constructional type of inductive voltage transformer depends on the value of the operating voltage and on its site of installation. For indoor applications measuring transformers usually have epoxy resin insulation, where as a rule all windings, and sometimes even the iron core, are embedded in the insulating material. The construction example of a single-pole insulated 10 kV voltage transformer is shown in Fig. 3.5-7.

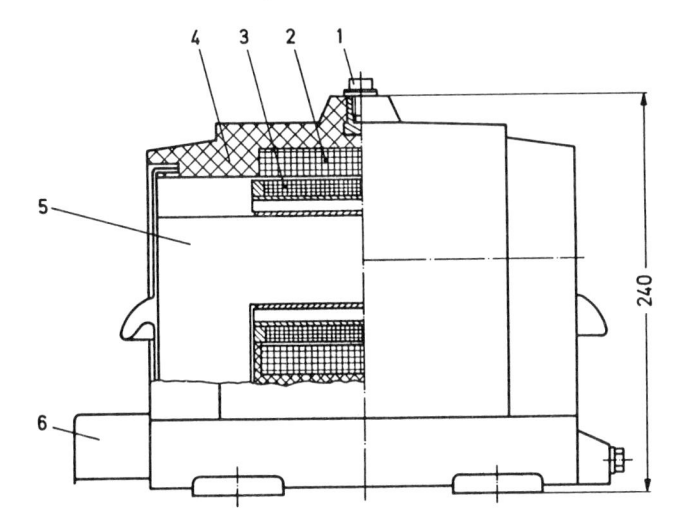

Fig. 3.5-7
Single-pole insulated epoxy resin voltage transformer for 10 kV (AEG)
1 primary terminal
2 primary winding
3 secondary windings (measuring winding and auxiliary winding for earth fault recognition)
4 epoxy resin body
5 core
6 secondary terminal box

For outdoor operation epoxy resin transformers have as yet only been introduced for not too high rated voltages. Single-pole transformers with oil-impregnated paper insulation find application here up to the highest voltages. Depending upon the arrangement of the active parts (cores and windings), one differentiates between a metal tank type and an insulated housing type of design. The advantage of the insulated housing type of design lies in its smaller constructional height; but, in contrast to the metal tank design, one must use porcelain housings with larger diameter. As an example, Fig. 3.5-8 shows a 110 kV voltage transformer in metal tank construction with metal bellows to take up the temperature dependent oil expansion [Hermstein 1959]. Fig. 3.5-9 represents a 800 kV voltage transformer in the insulated housing design. A peculiarity of this transformer is its construction as a four-stage cascade, where each coil is designed for only a quarter of the total voltage and the core lies at mid-potential. For very high voltages the cascade connection offers the possibility of subdividing the transformer into individually transportable stages.

An important special case are voltage transformers for metal encapsulated SF_6-insulated setups; they are built either with epoxy resin insulation or with gas-impregnated foil insulation. Fig. 3.5-10 shows an epoxy resin insulated voltage transformer for an SF_6-insulated switching station.

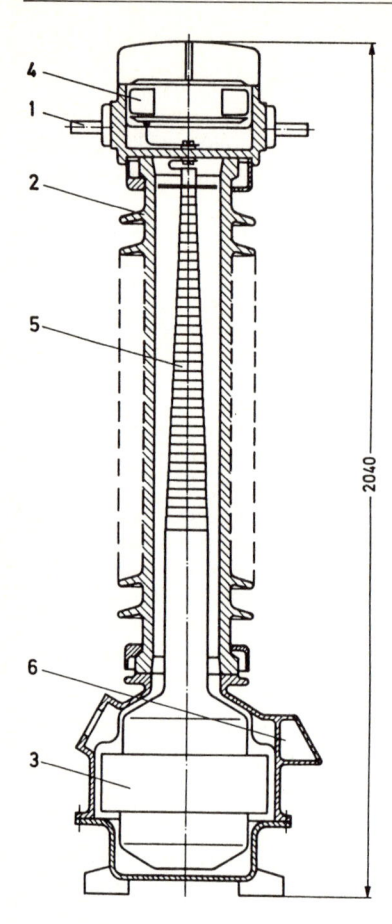

Fig. 3.5-8 110 kV potential transformer in metal tank design (Siemens)
1 primary terminal
2 porcelain housing
3 core with windings
4 bellows
5 capacitor grading
6 secondary terminal box

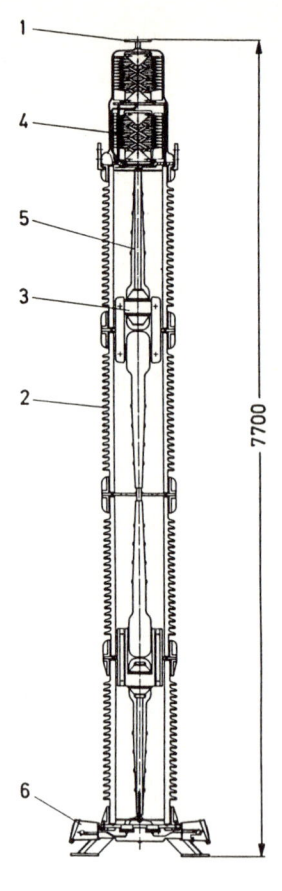

Fig. 3.5-9 800 kV voltage transformer in cascade connection (Ritz)
1 primary terminal
2 porcelain housing
3 core with windings
4 bellows
5 lead-out
6 secondary terminal

3.5.2 Capacitive voltage transformers

At rated voltages greater than about 110 kV capacitive voltage transformers are often used instead of inductive voltage transformers, for economic reasons. Special features of these devices shall therefore be briefly discussed here.

a) Circuitry

A capacitive voltage divider can be represented as an active two-pole with capacitive internal resistance referred to the secondary terminals (Fig. 3.5-11). On application as a

3.5 Instrument transformers

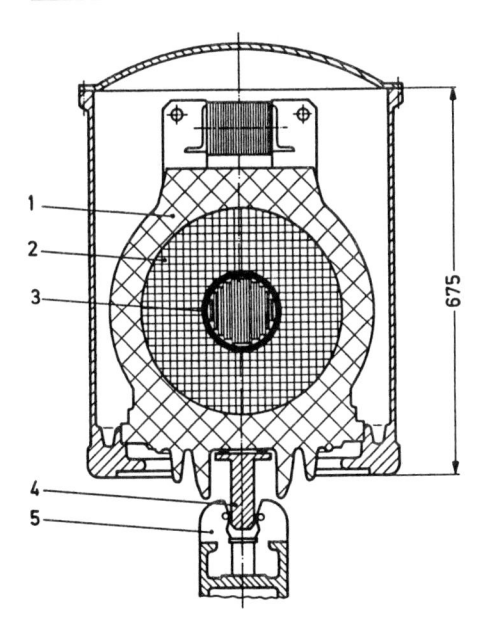

Fig. 3.5-10

Voltage transformer for SF_6-insulated 110 kV substation (MWB)

1 epoxy resin insulation
2 primary winding
3 secondary winding
4 plug connector
5 tulip contact

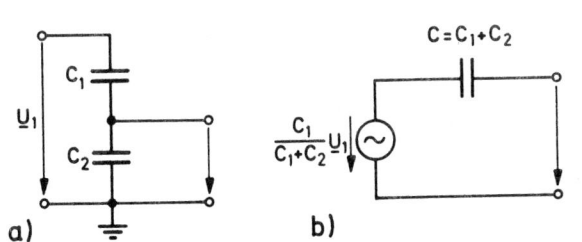

Fig. 3.5-11

Capacitive voltage divider
a) circuit
b) equivalent circuit

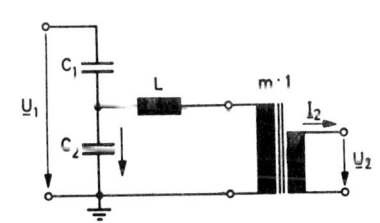

Fig. 3.5-12

Circuit diagram of a capacitive voltage transformer

voltage transformer the burden to be connected would have an impermissible feedback on the secondary voltage. The capacitive internal resistance can, however, be compensated by the series connection of a resonant inductance for a definite measuring frequency; the performance of capacitive voltage transformers is based upon this principle.

As will be shown futher on, pure capacitive division of high-voltages down to the standard secondary voltages is uneconomical. Therefore, in capacitive transformers one uses an inductive intermediate transformer in series connection. The primary voltage is initially subdivided capacitively to about 10kV to 30kV and then transformed down to the standard secondary voltage (Fig. 3.5-12). With regard to the capacitive divider, these can only be in single-pole construction. An earth fault winding can also be provided in exactly the same way as in inductive transformers.

b) Error calculation

The equivalent circuit of the capacitive voltage transformer, referred to the secondary side of the intermediate transformer, is shown in Fig. 3.5-13.

The voltage $\underline{U}'_1$ is obtained from the transformation ratio of the capacitive divider and the transformation ratio m of the inductive intermediate transformer as

$$\underline{U}'_1 = \underline{U}_1 \frac{C_1}{C_1 + C_2} \frac{1}{m}.$$

Further, with $C = C_1 + C_2$, we have:

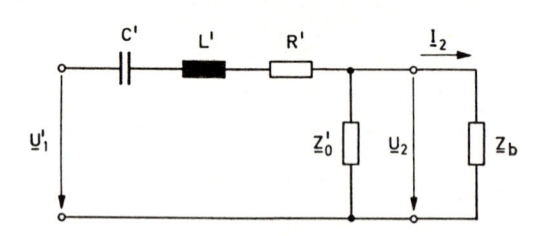

$$C' = m^2 C \quad \text{and} \quad L' = L/m^2.$$

Fig. 3.5-13 Equivalent circuit of a capacitive voltage transformer (as an approximation, $\underline{Z}'_0$ is assumed parallel to $\underline{Z}_b$)

The inductance L is equal to the sum of the inductance of the resonance coil and the short-circuit inductance of the intermediate transformer. R' and $\underline{Z}'_0$ are the effective resistance and the parallel impedance determined by the magnetizing current of the intermediate transformer, both referred to the secondary terminals.

The error $\underline{F}$ of the capacitive voltage transformer, as in an inductive voltage transformer, consists of the sum of the open-circuit error $\underline{F}_0$ and the loading error $\underline{F}_b$. Since only the last-named component represents a peculiarity in capacitive transformers, this alone shall be discussed here. From the definition of the error, we obtain from Fig. 3.5-13:

$$\underline{F}_b = -\frac{\underline{I}_2}{\underline{U}'_1}\left(R' + j\omega L' + \frac{1}{j\omega C'}\right) \approx -\frac{R' + j\left(\omega L' - \frac{1}{\omega C'}\right)}{\underline{Z}_b}.$$

When the error is calculated, deviations $\Delta\omega$ from the rated frequency ω_0 must be taken into account:

$$\omega = \omega_0 + \Delta\omega, \quad \text{where } \Delta\omega \ll \omega_0.$$

Then the following expressions result:

$$\omega L' = \omega_0 L' + \Delta\omega L'$$

$$\frac{1}{\omega C'} = \frac{1}{(\omega_0 + \Delta\omega)C'} \approx \frac{1}{\omega_0 C'}\left(1 - \frac{\Delta\omega}{\omega_0}\right).$$

The inductance L is now chosen so that at rated frequency the reactive component of the loading error is just compensated. The balance condition then reads:

$$\omega_0 L = \frac{1}{\omega_0 C}.$$

3.5 Instrument transformers

Finally for $\underline{F}_b$ we have:

$$\underline{F}_b = \frac{R' + j2\Delta\omega L'}{\underline{Z}_b}.$$

Both components of the loading error arise due to ohmic losses in the circuit and the assumed frequency deviation. But resonance mis-matching can also occur as a consequence of the deviations of the inductance L and the capacitance C from the corresponding nominal values. In capacitive transformers the adherence to certain error limits is therefore specified for a certain frequency range.

As a measure of the cost of capacitive dividers as in Fig. 3.5-11a we can take its reactive power rating:

$$P_b = U_1^2 \omega_0 \frac{C_1 C_2}{C_1 + C_2}.$$

Using the equivalent circuit given in Fig. 3.5-13, for the maximum frequency dependent imaginary component of the loading error we obtain the expression:

$$F_{b\delta} = -\frac{2\Delta\omega}{\omega_0} \frac{U_2^2}{Z_b} \frac{1}{P_b} \frac{C_2}{C_1}.$$

This relationship clearly shows the effect of P_b as well as of the division of the total voltage between the capacitive divider and the intermediate transformer. A small frequency dependence necessitates high reactive power and high tapping voltage.

The error diagram (Fig. 3.5-5) derived for the inductive voltage transformer, as well as the statements made in Section 3.5.1b about the determination of the open-circuit point, are also valid for capacitive voltage transformers.

c) Designs

The usual design of a capacitive voltage transformer for high voltages is shown in Fig. 3.5-14. The series connected capacitor windings with oil-impregnated paper dielectric are arranged in a slim porcelain housing. The resonance coil and the inductive intermediate transformer are housed in a metal chamber. The terminal X can either be directly earthed or serve to couple carrier frequency systems to the power line. In this case the series connection of C_1 and C_2, effective as coupling capacitance, must possess a certain minimum value of e.g. 4400 pF.

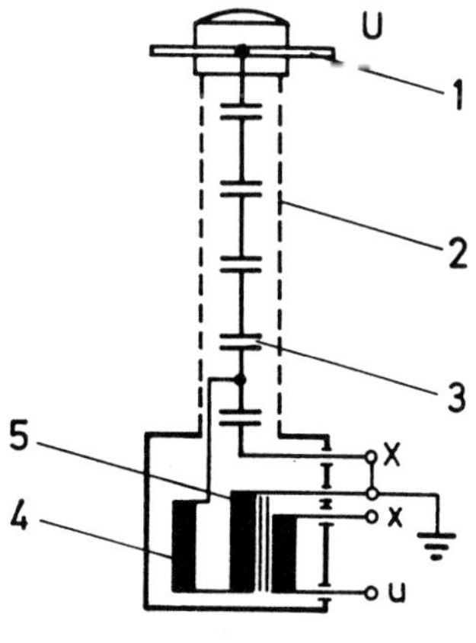

Fig. 3.5-14
Basic construction of a capacitive voltage transformer
1 primary terminal
2 porcelain housing
3 high-voltage capacitor
4 resonance coil
5 intermediate transformer

Capacitive voltage transformers are as a rule cheaper than inductive voltage transformers for high rated voltages greater than about 110kV, particularly if they are used simultaneously as coupling capacitors. Decisive disadvantages, however, are primarily due to the non-linear main inductance of the intermediate transformer. Ferro-resonant oscillations can lead to appreciable over-voltages and to impermissible heating of the magnetic cores and windings. To damp these oscillations damping elements are necessary, which should not however affect the true response of the measuring systems even for very fast events [*Gertsch* 1960; *Rosenberger* 1966]. In addition, compensating elements are necessary since the capacitance of the capacitors cannot be determined in advance as accurately as the instrument transformer accuracy demands.

Finally, one should also point out that during momentary breaking of a faulty network the line with only capacitive voltage transformers retains its static charge. On remaking the circuit this can result in high switching over-voltages. In contrast, inductive voltage transformers can discharge the line without risk. One therefore uses inductive voltage transformers, even for high voltages, although capacitive voltage transformers represent the more economical high-voltage technical solution by virtue of the simple construction of their insulation.

3.5.3 Current transformers

The operation of current transformers almost corresponds to that of a short-circuited transformer. The working range in current transformers for measuring purposes often extends from 0.05 to 1.2 times the measuring current; current transformers for protection purposes must usually be capable of working at up to more than 10 times the rated current.

a) Circuitry

Fig. 3.5-15 shows the circuit symbol of a current transformer with two cores as well as the corresponding circuit diagram. Current transformers are frequently provided with several cores each of which carries a secondary winding and is excited from a common primary winding. In this way the properties of the cores can be optimally matched to suit the different purposes of measurement and protection.

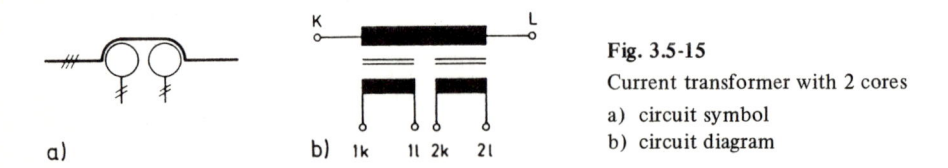

Fig. 3.5-15
Current transformer with 2 cores
a) circuit symbol
b) circuit diagram

In order to be able to use a given current transformer over a larger range of rated currents, the primary winding is often subdivided into a large number of groups connected in series or in parallel. The rated transformation ratio can then vary, for example, as 1:2:4, yet the rated current linkage and with it the error performance remain unchanged.

3.5 Instrument transformers

b) Error calculation

For calculation of the error of current transformers the transformer equivalent circuit of Fig. 3.5-16 is useful.

The error $\underline{F}$ is defined as:

$$\underline{F} = \frac{\underline{I}'_2 - \underline{I}_1}{\underline{I}_1} = \frac{\underline{I}_2 - \underline{I}'_1}{\underline{I}'_1}.$$

Taking the parameters of the equivalent circuit and neglecting errors of second order we have:

$$\underline{F} = -\frac{\underline{I}'_0}{\underline{I}'_1} \approx -\frac{\underline{I}'_0}{\underline{I}_2}.$$

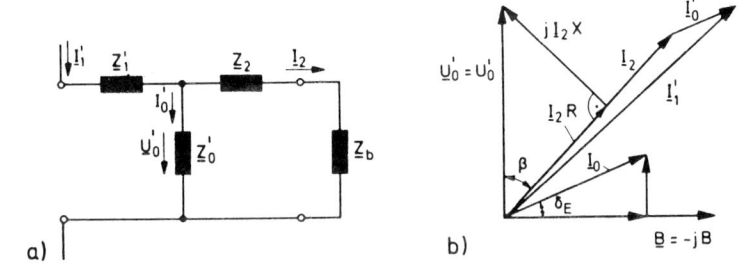

Fig. 3.5-16 Current transformer. a) equivalent circuit, b) phasor diagram

Thus the error is equal to the magnetizing current referred to the primary current. Except for the special types for protection purposes, current transformers are constructed with a closed iron core. On account of the non-linearity of the magnetization curve, the error is also equally non-linear with respect to Z_b; and so a simple error diagram cannot be developed for current transformers as it can for the case of voltage transformers.

When the secondary terminals are opened the normally small magnetizing current becomes equal to the primary current and the core is saturated. At the open secondary terminals a greatly distorted high alternating voltage appears which, on account of the rapid time-variation of the magnetic flux shows high-voltage peaks during current zero. Therefore current transformers must always be operated with a low ohmic burden or in short-circuit.

To calculate the error the magnetizing current is referred back to the design parameters of the circuit. Fig. 3.5-16a gives:

$$\underline{I}_2 = \frac{\underline{U}'_0}{\underline{Z}_2 + \underline{Z}_b}$$

with

$$\underline{U}'_0 = j\omega w_2 q_E \underline{B}$$

according to the law of induction.

From the law of magnetomotive force it follows:

$$\underline{I}'_0 = \frac{\underline{H} l_E}{w_2} = \frac{l_E \underline{B}}{w_2 \mu_0 \mu_r}.$$

When drawing the phasor diagram (Fig. 3.5-16b), it is expedient to keep the induced voltage in the direction of the real axis ($\underline{U}_0 = U_0$). Then we have:

$$\underline{I}_0 = I_0 (\sin \delta_E - j \cos \delta_E)$$

mit

$$I_0 = \frac{U_0 \, l_E}{\omega w_1^2 \, q_E \, \mu_0 \mu_r}$$

or

$$I_0' = \frac{U_0' \, l_E}{\omega w_2^2 \, q_E \, \mu_0 \mu_r} \,.$$

If the abbreviation

$$\underline{Z}_2 + \underline{Z}_b = \underline{Z} = R + jX$$

is introduced, on substitution we obtain:

$$\underline{F} = - \frac{l_E}{\omega w_2^2 \, q_E \, \mu_0 \mu_r} (\sin \delta_E - j \cos \delta_E)(R + jX) \,.$$

The error can once again be divided into a current error F_i and a phase angle error $F_{\delta i}$:

$$\underline{F} = F_i + j F_{\delta i} \,.$$

For both error components it follows:

$$F_i = - \frac{l_E}{\omega w_2^2 \, q_E \, \mu_0 \mu_r} (R \sin \delta_E + X \cos \delta_E)$$

$$F_{\delta i} = - \frac{l_E}{\omega w_2^2 \, q_E \, \mu_0 \mu_r} (X \sin \delta_E - R \cos \delta_E) \,.$$

The modulus of the error is

$$F = \frac{l_E}{\omega w_2^2 \, q_E \, \mu_0 \mu_r} Z \,.$$

From this some important relationships can be recognized which are significant for the design of the current transformer. The error is proportional to the length of the iron path l_E and inversely proportional to the iron cross-section q_E. An increase in l_E which would be desirable to reduce the electric field strength in the insulation requires a quadratic increase in core volume, if the error is to remain unchanged. It is therefore necessary to aim for a very compact type of construction in order to obtain as small lengths of iron as possible. The most important influencing parameter is, however, the rated current linkage Θ_n of the transformer. We have:

$$\Theta_n = I_{1n} w_1 \approx I_{2n} w_2 \,.$$

Since the secondary rated current is a fixed parameter the change in the error is inversely proportional to the square of the rated current linkage. This shows that one must choose the rated current linkage and hence, for a given primary rated current, the primary number of turns as large as possible to keep the error small. However, during a short-circuit this leads to large dynamic stress due to the magnetic forces in the windings.

3.5 Instrument transformers

In a practical calculation one must proceed stepwise and first determine approximately the induction which is proportional to the current, using given values of Z_b and $\cos\beta$, neglecting Z_2 and also with an assumption of w_2. From this we then obtain, for the chosen kind of iron via $\mu_r = f(B)$, a point-by-point calculation of the error. Having determined the main data, a complete calculation can then finally be made.

For current transformers of high accuracy one preferably chooses nickel-iron alloys, which already show high permeability even at low induction values. Disadvantageous, however, is the steep drop in permeability at relatively low induction. In current transformers for protection purposes, in which accuracy takes second place in favour of an extended measuring range, one uses silicon-iron alloys instead which have a high saturating induction. In Appendix 4 magnetic properties required for dimensioning current transformers are compiled for typical core materials.

c) Performance under overcurrent

The overcurrent performance of current transformers is described by the overcurrent factor n. The magnitude of the primary current, for which certain error conditions must still be satisfied, is called the error limit current nI_{1n}. Fig. 3.5-17 shows the dependence of the secondary current $I'_2 = I_2 w_2/w_1$ (referred to w_1) upon the primary current I_1, where only the moduli of the currents are represented. An ideal transformer would correspond to the dashed straight line, whilst the real transformer has the continuously drawn characteristic because of the iron saturation. For measurement purposes one usually operates in the linear region of this curve. Protective current transformers should only go into saturation at a later stage; they have a high overcurrent factor e.g. n = 10; current transformers for service measurements, on the other hand should go into saturation earlier to protect the connected equipment against overload; they therefore have a low overcurrent factor, e.g. n = 5.

Actually in the overcurrent region the time curve of the secondary current of current transformers with closed iron core deviates appreciably from the sinusoidal form. The effective magnetizing current referred to the primary current is denoted the total error F_g:

$$F_g = \frac{1}{I_1} \sqrt{\frac{1}{T} \int_0^T (i'_2 - i_1)^2 \, dt} \, .$$

Here i_2 and i_1 are the instantaneous values of the currents and T is the periodic time. Approximately only, we have:

$$F_g \approx \left| \frac{I'_2 - I_1}{I_1} \right| = |\underline{F}| \, .$$

Fig. 3.5-17
Current transformer under overcurrent

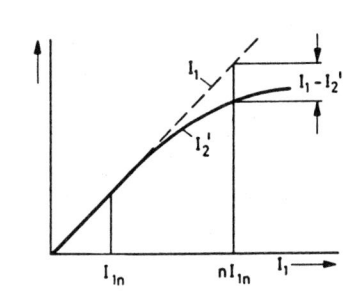

156   3 Design and Manufacture of High-Voltage Equipment

According to existing standards for measuring cores (M = measuring) the error limit current $F_g \geq 15\%$; the class designation 0.5/M5 signifies, for example, that the error limits correspond to 0.5% and n = 5. In protection cores (P = protection), the designation 10P 20 signifies, for example, that for n = 20, the total error F_g must be $\leq 10\%$.

d) Designs

Corresponding to the number of primary windings, one differentiates between single conductor current transformers and wound current transformers. The design with the primary winding as a straight bar conductor results in a specially simple, short-circuit proof construction; however, it has the disadvantage that higher measuring efficiency for economical core dimensions can usually be obtained only at rated currents greater than about 1000 A.

As in inductive voltage transformers even for current transformers the construction with epoxy resin insulation dominates for indoor applications up to 110 kV rated voltage. It has supplanted the earlier commonly adopted construction using porcelain in the medium voltage range due to the higher short-circuit strength of the windings which are completely embedded in the insulating material.

According to the design, see Fig. 3.5-18, one differentiates between support and bushing current transformers. A section through a 10 kV support current transformer is reproduced in Fig. 3.5-19.

Fig. 3.5-18
Designs of epoxy resin impregnated current transformers
a) support current transformer as a wound current transformer
b) bushing current transformer as a single conductor current transformer
1 core
2 primary winding
3 secondary winding

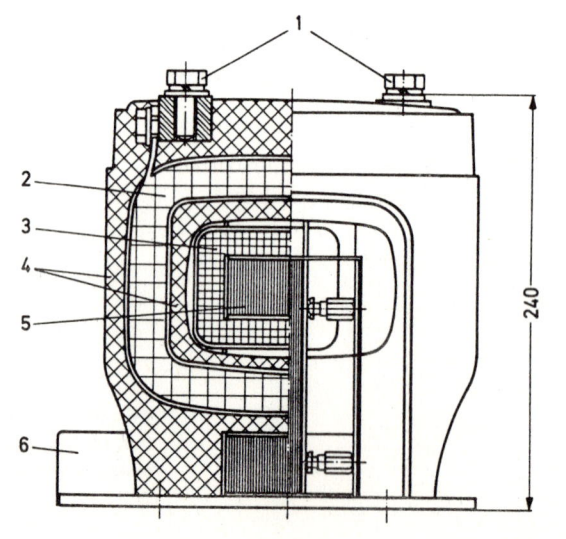

Fig. 3.5-19
Support current transformer with epoxy resin insulation for 10 kV (AEG)
1 primary terminal
2 primary winding
3 secondary windings
4 epoxy resin body
5 core
6 secondary terminal box

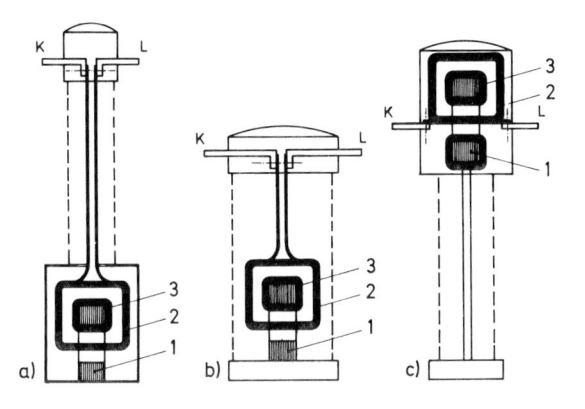

3.5 Instrument transformers

Fig. 3.5-20
Design of outdoor current transformers
a) metal tank type
b) insulating cylinder type
c) overhead type
1 core, 2 primary winding, 3 secondary winding

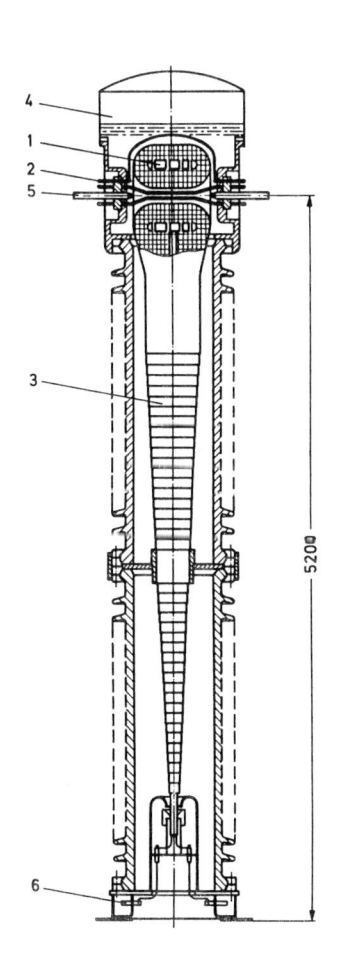

Fig. 3.5-21
400 kV overhead current transformer (MWB)
1 cores
2 re-connectable primary windings
3 capacitively graded secondary lead-out
4 nitrogen cushion
5 primary terminal
6 secondary terminal

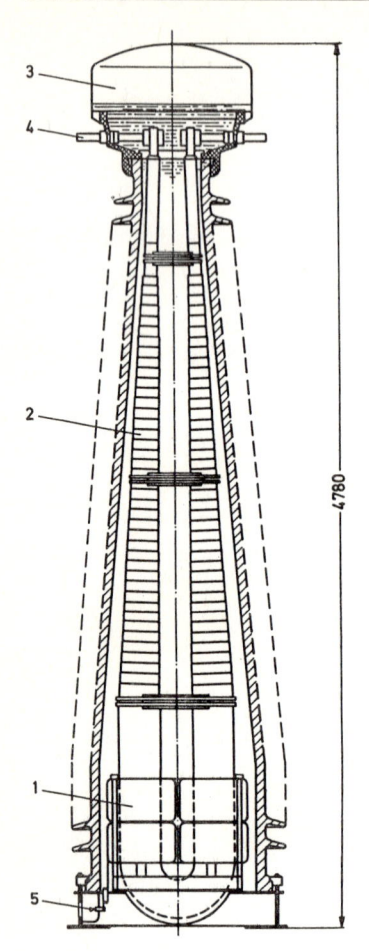

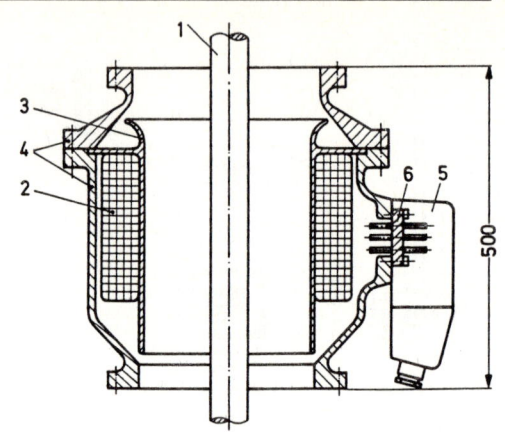

Fig. 3.5-23 Current transformer for SF_6-insulated 110 kV substation (BBC)
1 tube inner conductor as primary winding
2 ring cores with secondary windings
3 shielding electrode
4 tube outer conductor as housing
5 secondary terminal box
6 gas-tight bushing plate

Fig. 3.5-22
500 kV current transformer with a U-type primary conductor (MWB)
1 cores
2 capacitively graded primary lead-out
3 nitrogen cushion
4 primary terminal
5 secondary terminal

Current transformers for high-voltage outdoor stations are built with oil-impregnated paper insulation and porcelain housing. Depending upon the arrangement of the active parts (cores, windings), one differentiates between the metal tank, the insulated housing and the overhead types of construction (Fig. 3.5-20). Current transformers of the last kind have the advantage of shorter leads in the primary circuit for high rated and short-circuit currents, as may be seen in the 400 kV current transformer shown in Fig. 3.5-21. For high voltages and currents, by means of a U-type construction for the primary conductor, a particularly well arranged layout of the insulation can be achieved as in the example of the 500 kV current transformer shown in Fig. 3.5-22.

In coaxial insulation systems as in cables, transformer bushings or SF_6-insulated busbars, it is often possible to construct single conductor current transformers without special insulation. Here very often a ring core with the secondary winding uniformly wound on it is inserted over the insulation, whereby the outer earthing layer must be led out of the current transformer. Fig. 3.5-23 shows a ring core current transformer in a SF_6-insulated substation.

Appendix 1

Utilization factors for simple electrode configurations

For many configurations of spherical or cylindrical electrodes the utilization factor η according to *Schwaiger*, i.e. the ratio of mean to maximum field strength,

$$\eta = \frac{E_{mean}}{E_{max}}$$

can be written as a function of one or two quantities known as "geometrical characteristics". These quantities are:

$$p = \frac{s+r}{r} \quad \text{and} \quad q = \frac{R}{r} .$$

Here s is the gap spacing, r and R are the radii of the electrodes. For a particular case, with an applied voltage U at the electrodes, the maximum field strength is then calculated as:

$$E_{max} = \frac{U}{s\eta} .$$

Utilization factors for a few simple electrode configurations are listed in the following:

A 1.1 Spherical configurations

Table A 1.1: Utilization factors for different spherical configurations

p	q = 1	q = 1	q = ∞	q = p
1	1	1	1	1
1.5	0.850	0.834	0.732	0.667
2	0.732	0.660	0.563	0.500
3	0.563	0.428	0.372	0.333
4	0.449	0.308	0.276	0.250
5	0.372	0.238	0.218	0.200
6	0.318	0.193	0.178	0.167
7	0.276	0.163	0.152	0.143
8	0.244	0.140	0.133	0.125
9	0.218	0.123	0.117	0.111
10	0.197	–	0.105	0.100
15	0.133	–	–	–

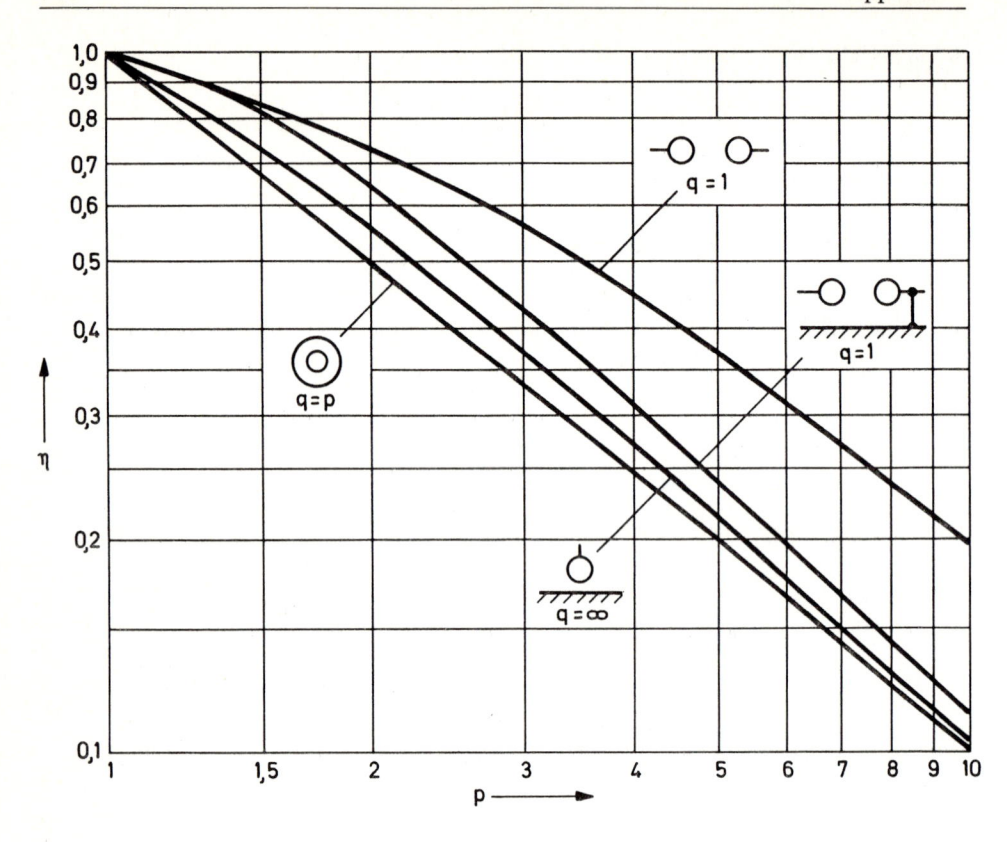

Fig. A 1.1-1 Utilization factors for different spherical configurations as a function of p (cf. Table A 1.1)

A 1.2 Cylindrical configurations

Table A 1.2: Utilization factors for different cylindrical configurations

p	q = 1	q = 2	q = 3	q = 5	q = 10	q = 20	q = ∞
1	1	1	1	1	1	1	1
1.5	0.924	0.894	0.884	0.878	0.871	0.864	0.861
2	0.861	0.815	0.798	0.783	0.772	0.766	0.760
3	0.760	0.702	0.679	0.658	0.641	0.632	0.623
4	0.684	0.623	0.595	0.574	0.555	0.548	0.533
5	0.623	0.564	0.538	0.513	0.492	0.486	0.468
6	0.574	0.517	0.488	0.469	0.450	0.435	0.419
8	0.497	0.447	0.420	0.401	0.377	0.368	0.349
10	0.442	0.397	0.375	0.352	0.330	0.324	0.301
15	0.349	0.314	0.296	0.277	0.257	0.249	0.228
20	0.291	0.263	0.248	0.232	0.214	0.202	0.186
50	0.1574	–	–	–	–	–	0.0932
100	0.094	–	–	–	–	–	0.0537
300	0.038	–	–	–	–	–	0.0214
500	0.025	–	–	–	–	–	0.0138
800	0.0168	–	–	–	–	–	0.00922
1000	0.0138	–	–	–	–	–	0.0076

p	q = p	q = 3	q = 5	q = 10	q = 20
1	1	1	1	1	1
1.5	0.811	0.831	0.847	0.855	0.857
2	0.693	0.717	0.735	0.748	0.754
3	0.549	0.549	0.582	0.604	0.614
4	0.462	–	0.478	0.507	0.521
5	0.402	–	0.402	0.439	0.454
6	0.358	–	–	0.386	0.404
8	0.297	–	–	0.310	0.331
10	0.256	–	–	0.256	0.281
15	0.193	–	–	–	0.204
20	0.158	–	–	–	0.158
50	0.0798	–	–	–	–
100	0.047	–	–	–	–
300	0.019	–	–	–	–
500	0.0125	–	–	–	–
800	0.0084	–	–	–	–
1000	0.0069	–	–	–	–

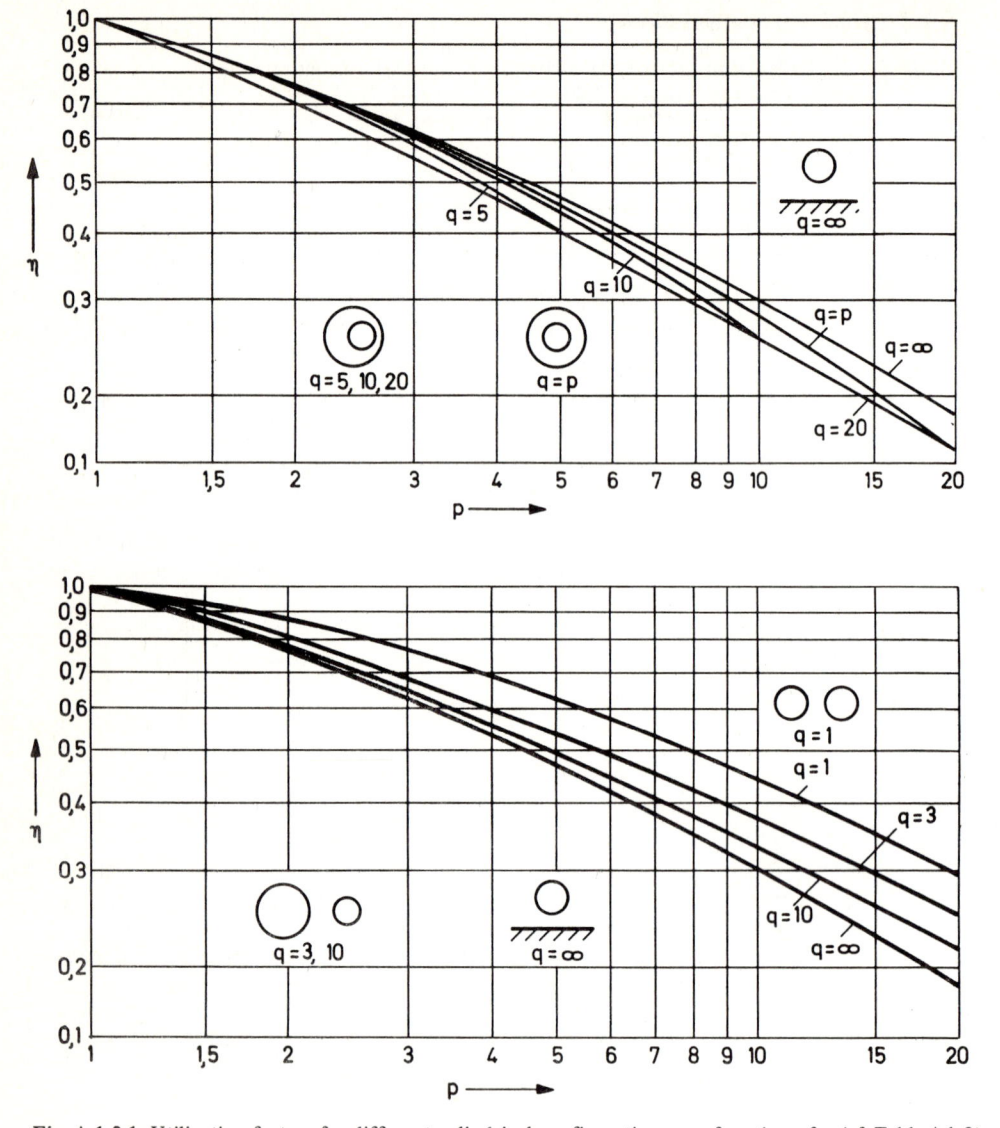

Fig. A 1.2-1 Utilization factors for different cylindrical configurations as a function of p (cf. Table A 1.2)

Utilization factors for simple electrode configurations 163

A 1.3 Point and knife-edge configurations

	Rotational configurations points		Translatory configurations knife-edges		
Electrode configuration	R1	R2	T1	T2	T3
Designation	confocal paraboloids	hyperboloid against plane	rounded right angular knife-edge against plane	confocal parabolic cylinder	hyperbolic knife-edge against plane
Utilization factor $\eta = f(p)$ see diagram	$\eta = \dfrac{\ln(2p-1)}{2(p-1)}$	$\eta = \dfrac{\ln(\sqrt{p}+\sqrt{p-1})}{\sqrt{p(p-1)}}$		$\eta = \dfrac{2}{1+\sqrt{2p-1}}$	$\eta = \dfrac{\operatorname{arccot}\dfrac{1}{\sqrt{p-1}}}{\sqrt{p-1}}$

Fig. A 1.3-1 Geometry and utilization factor of rotational point configurations and translatory knife-edge configurations

Bibliography: [*Prinz, Singer* 1967; *Vajda* 1966; *Dreyfus* 1924]

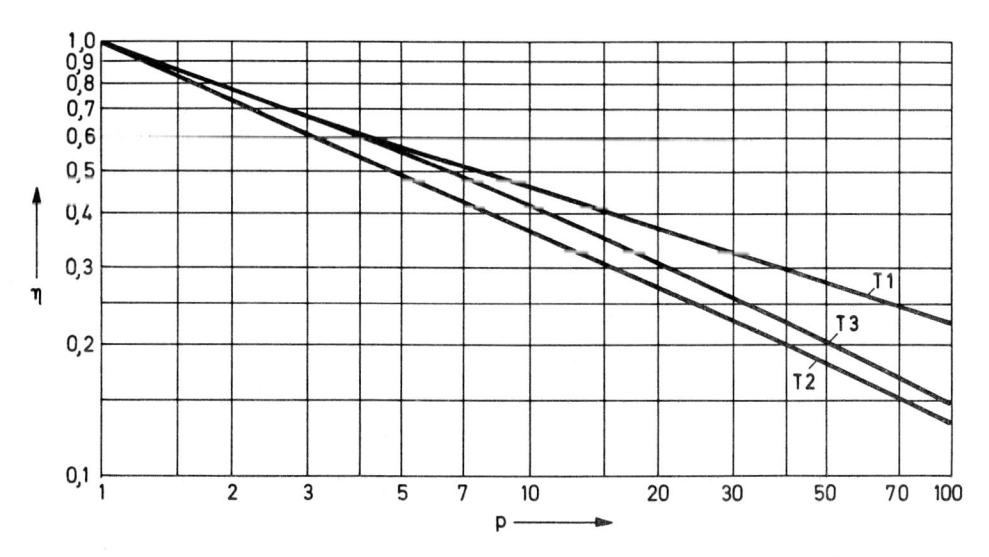

Fig. A 1.3-2 Utilization factors for translatory knife-edge configurations as a function of p (cf. Fig. A 1.3-1)

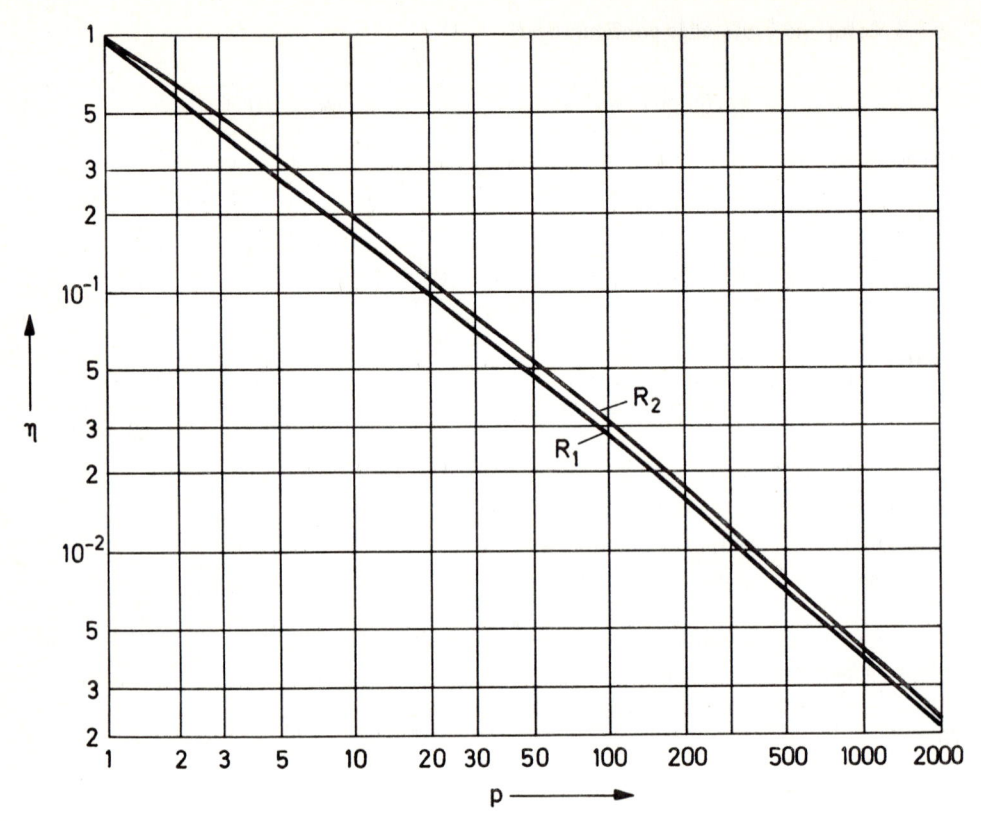

Fig. A 1.3-3 Utilization factors for rotational point configurations as a function of p (cf. Fig. A 1.3-1)

A 1.4 Circular ring configurations

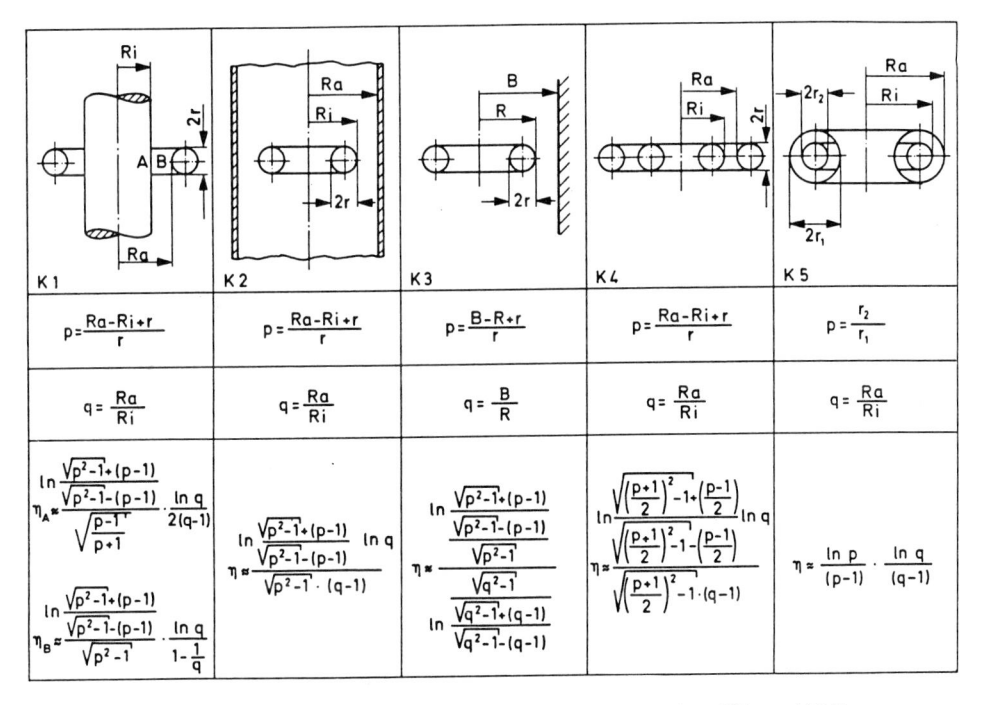

K 1	K 2	K 3	K 4	K 5
$p = \dfrac{Ra-Ri+r}{r}$	$p = \dfrac{Ra-Ri+r}{r}$	$p = \dfrac{B-R+r}{r}$	$p = \dfrac{Ra-Ri+r}{r}$	$p = \dfrac{r_2}{r_1}$
$q = \dfrac{Ra}{Ri}$	$q = \dfrac{Ra}{Ri}$	$q = \dfrac{B}{R}$	$q = \dfrac{Ra}{Ri}$	$q = \dfrac{Ra}{Ri}$
$\eta_A \approx \dfrac{\ln\dfrac{\sqrt{p^2-1}+(p-1)}{\sqrt{p^2-1}-(p-1)}}{\sqrt{\dfrac{p-1}{p+1}}} \cdot \dfrac{\ln q}{2(q-1)}$ $\eta_B \approx \dfrac{\ln\dfrac{\sqrt{p^2-1}+(p-1)}{\sqrt{p^2-1}-(p-1)}}{\sqrt{p^2-1}} \cdot \dfrac{\ln q}{1-\dfrac{1}{q}}$	$\eta \approx \dfrac{\ln\dfrac{\sqrt{p^2-1}+(p-1)}{\sqrt{p^2-1}-(p-1)}}{\sqrt{p^2-1}\cdot(q-1)} \ln q$	$\eta \approx \dfrac{\ln\dfrac{\sqrt{p^2-1}+(p-1)}{\sqrt{p^2-1}-(p-1)}}{\sqrt{p^2-1}} \cdot \dfrac{\ln\dfrac{\sqrt{q^2-1}+(q-1)}{\sqrt{q^2-1}-(q-1)}}{\sqrt{q^2-1}}$	$\eta \approx \dfrac{\ln\dfrac{\sqrt{\left(\dfrac{p+1}{2}\right)^2-1}+\left(\dfrac{p-1}{2}\right)}{\sqrt{\left(\dfrac{p+1}{2}\right)^2-1}-\left(\dfrac{p-1}{2}\right)}}{\sqrt{\left(\dfrac{p+1}{2}\right)^2-1}\cdot(q-1)} \ln q$	$\eta \approx \dfrac{\ln p}{(p-1)} \cdot \dfrac{\ln q}{(q-1)}$

Fig. A 1.4-1 Geometry and utilization factor of circular ring configurations [*Morva* 1966]
The maximum field strength can appear at A or B

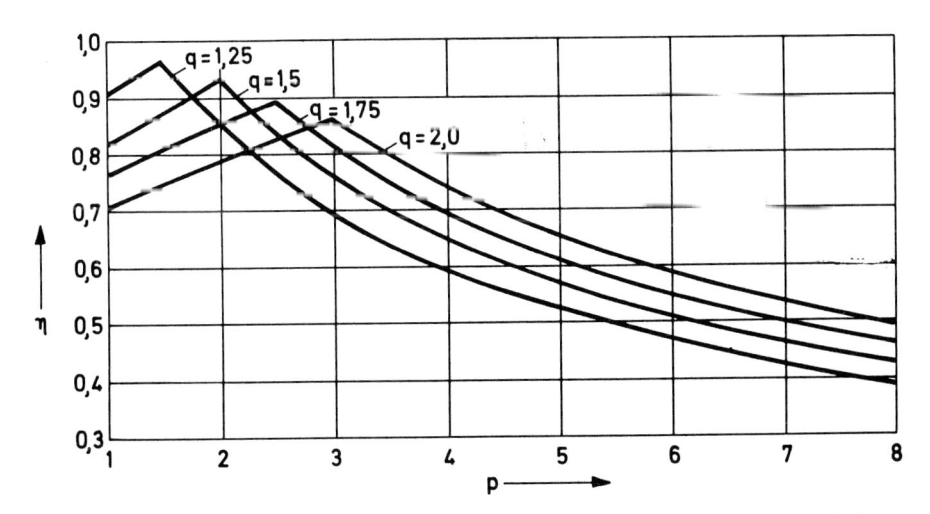

Fig. A 1.4-2 Utilization factors of a configuration comprising a circular ring round a cylinder as a function of p (configuration K 1 in Fig. A 1.4-1)

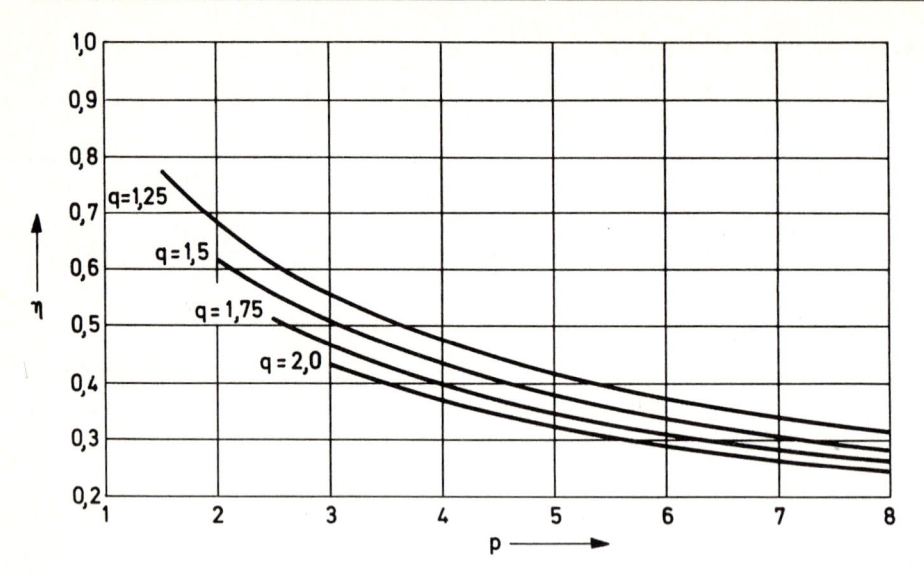

Fig. A 1.4-3 Utilization factors of a configuration comprising a circular ring inside a cylinder as a function of p (configuration K2 in Fig. A 1.4-1)

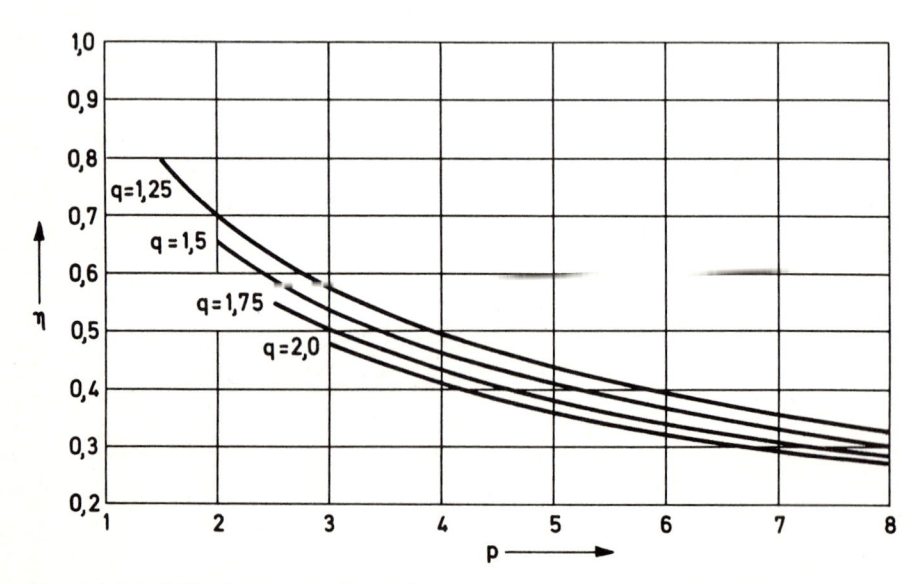

Fig. A 1.4-4 Utilization factors of a configuration comprising a circular ring and a plane as a function of p (configuration K3 in Fig. A 1.4-1)

Utilization factors for simple electrode configurations 167

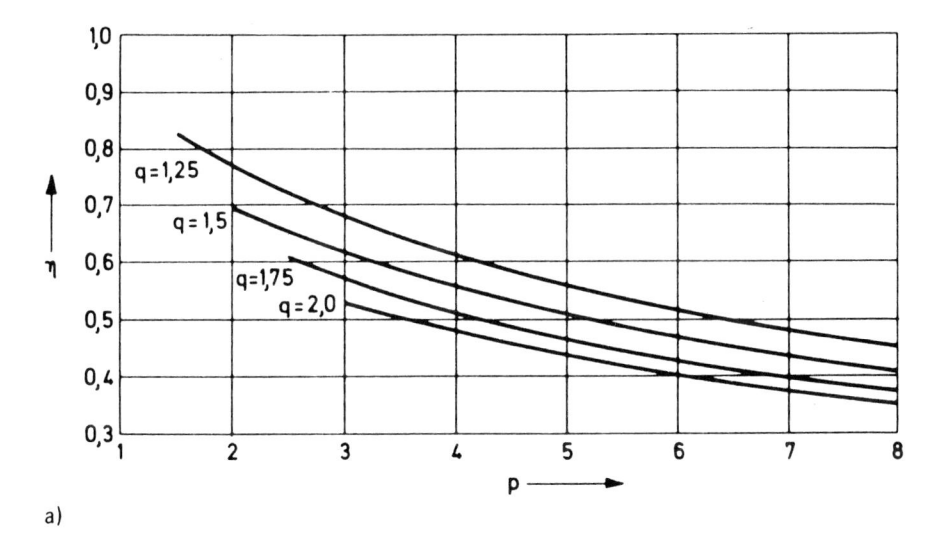

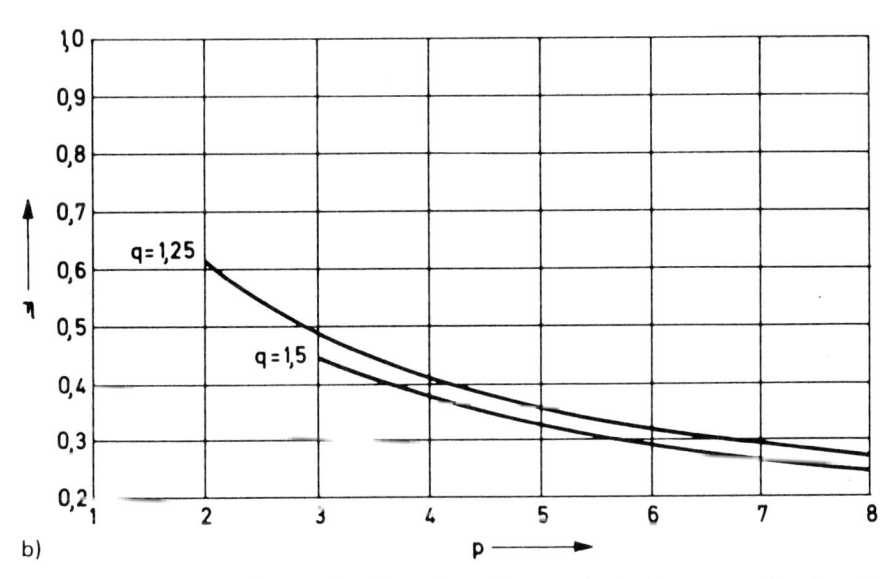

Fig. A 1.4-5 Utilization factors of configurations of concentric circular rings as a function of p
a) configuration K4 in Fig. A 1.4-1, b) configuration K5 in Fig. A 1.4-1

Appendix 2

Electric strength of gas-insulated configurations

A 2.1 Breakdown voltage in the homogeneous field

According to Section 1.2.2, the breakdown voltage $\hat{U}_d$ in a homogeneous field is constant if the product ps is held constant. This Paschen law is satisfied up to certain limits of the product ps. For all gases, the function $\hat{U}_d = f(ps)$ has a typical form with a distinct minimum. Paschen curves for air, SF_6, H_2 and Ne are represented in Fig. A 2.1-1 [*Gänger* 1953; *Dakin* et al. 1974; *Hess* 1976].

The curves are valid for 20 °C corresponding to $T_0 = 293$ K. For an absolute temperature T differing from that, the required breakdown voltage should be read at the abscissa value of $ps \frac{T_0}{T}$.

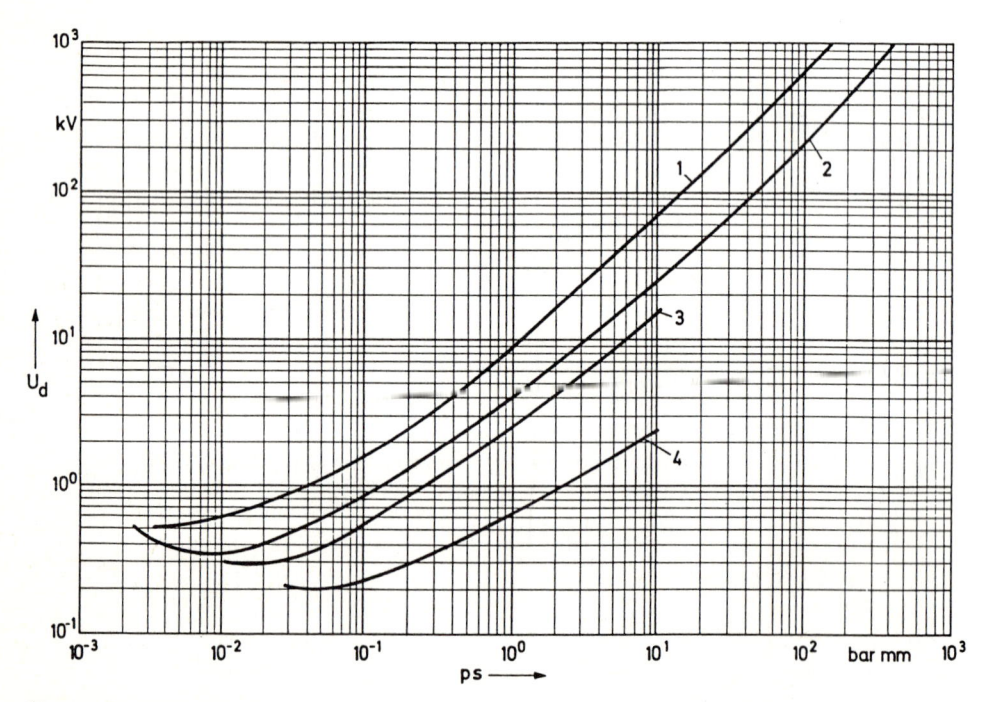

Fig. A 2.1-1 Paschen curves for various gases
1 SF_6, 2 Air ($\approx N_2$), 3 H_2, 4 Ne

Electric strength of gas-insulated configurations

A 2.2 Breakdown field strengths of plate, cylinder and sphere electrodes

The breakdown field strength $\hat{E}_d$ of gases depends not only upon temperature, pressure and type of gas but also upon the geometrical dimensions of the electrodes. For practical application it is generally sufficient to take the geometry into account by comparison with simple configurations. Therefore, in the following the breakdown field strengths of plate, cylinder and sphere electrodes are listed for various gases and geometries [*Schumann* 1923; *Nitta, Shibuya* 1971; *Mosch, Hauschild* 1978].

The diagrams are valid for the pressure $p_0 = 1013$ mbar and absolute temperature $T_0 = 293$ K. For different values of p and T, in the range of 1...10 bar, we have approximately:

$$\hat{E}_d = \hat{E}_{d0} \left(\frac{p/p_0}{T/T_0}\right)^\alpha \text{ with } \alpha = 0.7...0.8 .$$

In technical configurations at high pressure, unevenness of the electrodes (sharp points, edges, grooves) and particles (dust, metal filings) effect a reduction of the breakdown voltage.

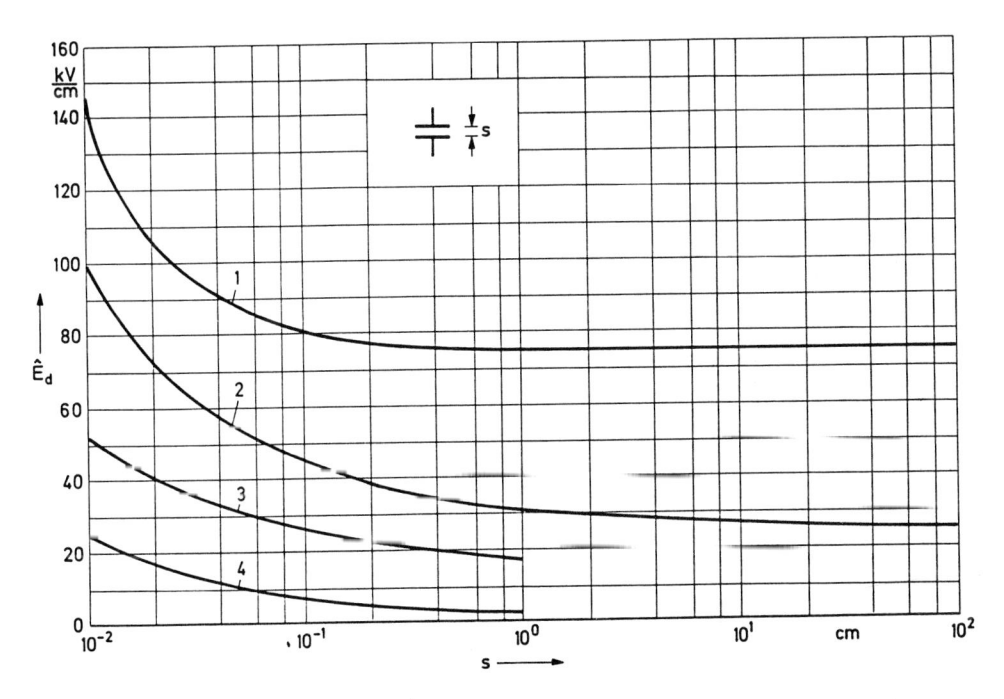

Fig. A 2.2-1 Breakdown field strength $\hat{E}_d$ of various gases in the homogeneous field of plate electrodes at 20 °C and 1013 mbar
1 SF_6, 2 Air, 3 H_2, 4 Ne

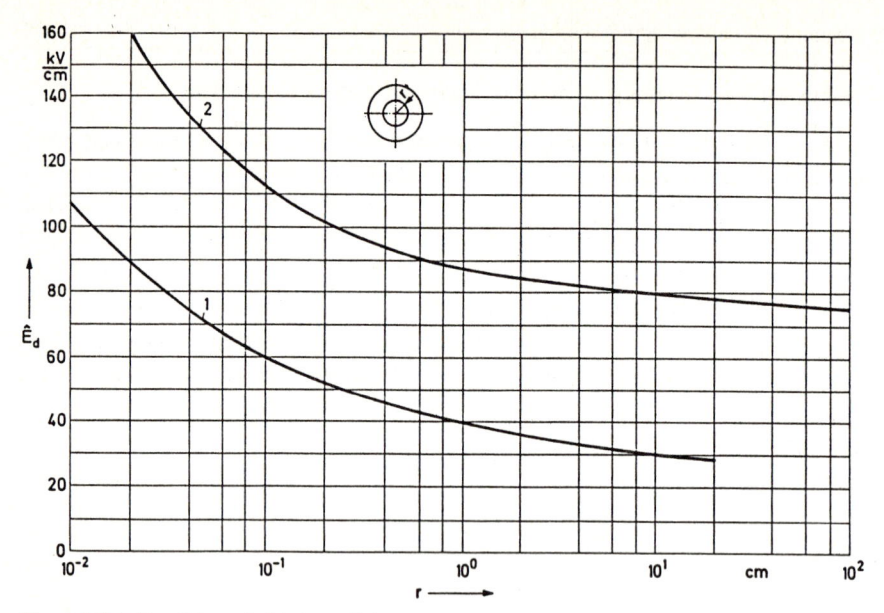

Fig. A 2.2-2 Breakdown field strength between electrodes with cylindrical surfaces at 20 °C and 1013 mbar. The values hold for parallel cylinders (enclosed, adjacent), or crossed cylinders, where for unequal cylinders r is the radius of the smaller cylinder.   1 Air,   2 SF_6

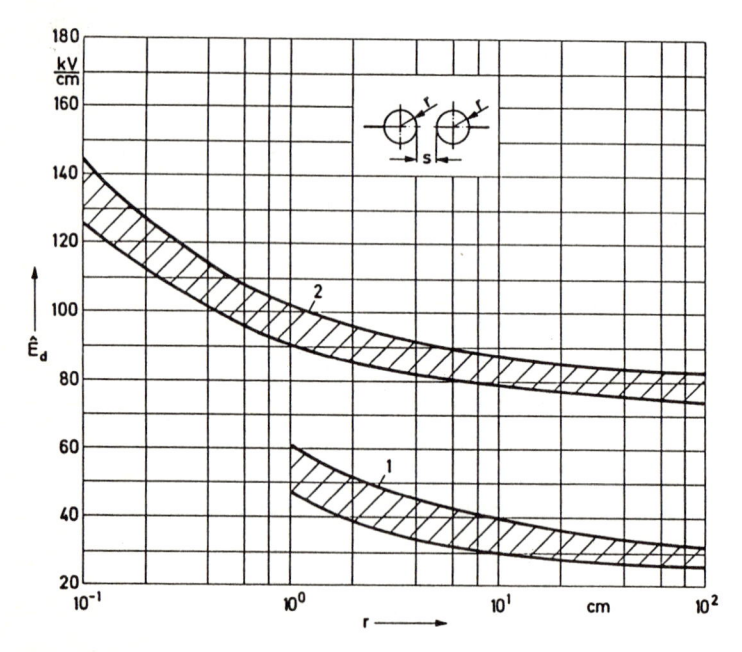

Fig. A 2.2-3 Breakdown field strength between electrodes with spherical surfaces in the range $s/r = 0.1...1$ at 20 °C and 1013 mbar. 1 Air, 2 SF_6

In a weakly inhomogeneous field, the breakdown field strength $\hat{E}_d$ depends upon the sphere radius r and the gap spacing s. In the representation $\hat{E}_d = f(r)$ with s as parameter, series of curves appear lying within the hatched region in the diagram for $s/r = 0.1...1$.

A 2.3 Breakdown voltage of rod gaps in air

The breakdown behaviour of large gap spacings in a strongly inhomogeneous field determines the space requirement of outdoor stations. Since the inception voltage $\hat{U}_e$ in these cases lies appreciably below the breakdown voltage $\hat{U}_d$, the shape of the electrodes is less significant. On the other hand, the voltage stress as a function of time has a significant effect on $\hat{U}_d$ as a consequence of long formative times.

Rod gaps show about the same qualitative and quantitative breakdown behaviour as other configurations with a strongly inhomogeneous field at the same gap spacing s. The relationship $\hat{U}_d = f(s)$ for rod gaps is therefore suited for the design of comparable air clearances.

The values are valid for standard atmospheric conditions (1013 mbar, 20 °C). For conversion to other values of pressure p and absolute temperature T, we have, approximately:

$$U_d \sim p/T .$$

The effect of the humidity of air can be approximately taken into account by a correction factor as in IEC Publ. 60-1, the reference humidity is 11 g/m^3, the correction factor lies between 0.9 and 1.1.

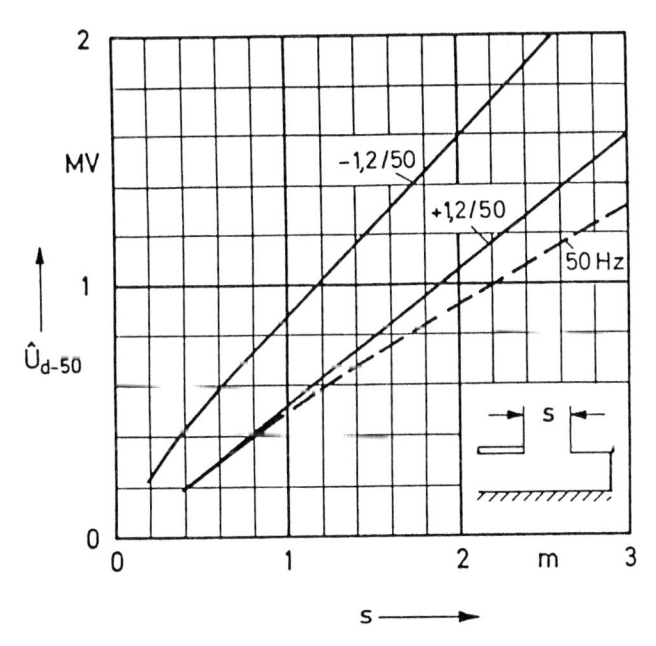

Fig. A 2.3-1 Breakdown voltage $\hat{U}_{d-50}$ of a rod-rod configuration for different types of voltage

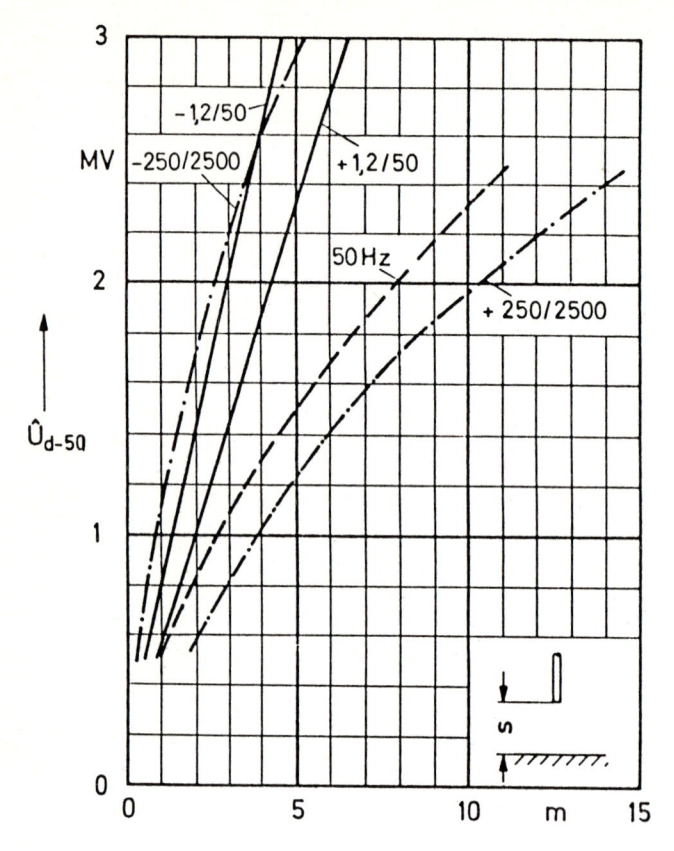

Fig. A 2.3-2 Breakdown voltage $\hat{U}_{d-50}$ of a rod-plate configuration for different types of voltage

Appendix 3

Properties of insulating materials

The following tables compile the essential physical properties of a number of insulating materials which are important to high-voltage technology. The numerical values were obtained from [*Hütte* 1955; *Brinkmann* 1975; *Oburger* 1957; *Saure* 1979], as well as from manufacturers' data. Characteristic values for other materials not listed here may be obtained from the quoted sources.

Table A 3.1: Thermal properties of materials

	Specific heat c in J/kg K	Thermal conductivitity λ in W/K m	Linear thermal expansion in 10^{-6}/K
Copper	390	393	16.2
Aluminium	920	221	23.8
Porcelain	800	1.5 ... 2.5	4 ... 6
Glass	800	0.7 ... 1.1	3 ... 9
EP-unfilled	1350	0.20	65
EP-filled	2150	0.95	35
PE	2100	0.3 ... 0.4	150 ... 320
PVC	1000	0.17	70 ... 190
PTFE	1000	0.25	120
Water	4180	0.60	–
Transformer oil	1800	0.15	–
Air	1000	0.026	–
Hydrogen	14212	0.179	–
SF_6	633	0.019	–

Table A 3.2: Properties of gases for 20 °C, 1013 mbar

	Density γ in g/dm^3	Ionization voltage U_i in V	Breakdown field strength E_d at s = 1 cm in kV/cm	Specific heat c in J/kg K	Thermal conductivity λ in 10^{-3} W/K m	Boiling point in °C
Air	1.21	–	32	1000	25.6	–
N_2	1.17	15.8	33	1038	25.5	–195.8
O_2	1.33	12.8	29	915	26.0	–183.0
CO_2	1.84	13.7	29	820	16.0	– 78.5
H_2	0.08	15.4	19	14212	179	–252.8
He	0.17	24.6	10	5225	149	–269.0
Ne	0.84	21.6	2.9	1030	48.3	–246.0
Ar	1.66	15.8	6.5	523	17.5	–185.9
Kr	3.48	14.0	8	–	9.4	–152.9
Xe	5.50	12.1	–	–	5.5	–170.1
SF_6	6.15	15.9	89	633	18.8	– 63.8

Table A 3.3: Properties of inorganic insulating materials

Property	Unit	Glass	Glass fibre (E-glass)	Quartz porcelain KER 110.1	Clay porcelain KER 110.2	Steatite KER 220
Breakdown field strength E_d	kV/mm	10...50	–	20...40	20...40	25...40
Volume resistivity ρ	Ωcm	$10^{12}...10^{14}$	10^{13}	10^{12}	10^{12}	10^{12}
Dielectric constant ϵ_r	–	4.5...7	6	6	6	6
Dissipation factor (1 MHz) tan δ	–	$10^{-2}...10^{-3}$	10^{-3}	$5 \cdot 10^{-3}$	$5 \cdot 10^{-3}$	$5 \cdot 10^{-4}$
Tracking index	–	KA 3c	KA 3c	KA 3c	KA 3c	KA 3c
Density γ	g/cm^3	2.2...2.6	2.5	2.3...2.4	2.5...2.6	2.6...2.7
E-modulus	kN/mm^2	60...90	70	50...100	50...100	80...120
Bending strength	N/mm^2	30...120	–	60...100	100...140	120...150
Tensile strength	N/mm^2	50...100	2500	25...40	40...60	60...90
Compressive strength	N/mm^2	800	–	250...500	400...700	800...1000
Thermal conductivity λ	W/Km	0.7...1.1	1.1	1.5...2.5	1.5...2.5	2...4
Linear thermal expansion	10^{-6}/K	3...9	5	4...6	4...6	6...9
Specific heat c	J/kg K	700...800	800	800	800	800
Arc withstand index	–	L6	–	L6	L6	L6

Table 3.4: Properties of liquid insulating materials

Property	Unit	Mineral oil	Chlorinated diphenyls	Silicon oil
Breakdown field strength E_d	kV/mm	25	20	10
Volume resistivity ρ	Ωcm	10^{14}	10^{14}	10^{15}
Dielectric constant ϵ_r	–	2.2	5.5	2.8
Dissipation factor (1 MHz) tan δ	–	10^{-3}	10^{-3}	$2 \cdot 10^{-4}$
Density γ	g/cm^3	0.89	1.5	0.97
Thermal conductivity λ	W/Km	0.14	0.1	0.16
Specific heat c	J/kg K	1800	–	2000
Thermal stability limit	°C	90	–	150
Flash point	°C	> 130	–	> 300
Neutralization number (acidity)	mg KOH/g	< 0.3	–	–
Saponification number	mg KOH/g	< 0.6	–	–

Properties of insulating materials 175

Table A 3.5: Properties of thermoplasts

Property	Unit	High-pressure PE (LDPE)	Low-pressure PE (HDPE)	Cross-linked PE (XLPE)	Hard PVC	Soft PVC	PTFE
Breakdown field strength E_d	kV/mm	75	100	50	30	10	25
Volume resistivity ρ	Ωcm	$5 \cdot 10^{17}$	$5 \cdot 10^{17}$	10^{16}	10^{15}	10^{14}	10^{17}
Dielectric constant ϵ_r	–	–	–	2.3	2.4	2.3	5.5	2.1
Dissipation factor (1 MHz) tan δ	–	$2 \cdot 10^{-4}$	10^{-3}	10^{-3}	$2 \cdot 10^{-2}$	10^{-1}	10^{-4}
Tracking Index	–	KA 3b	KA 3c	KA 3c	KA 3a	KA 1	KA 3c
Density γ	g/cm^3	0.92	0.95	0.92	1.4	1.2...1.3	2.15
E-modulus	kN/mm^2	0.15	0.7	0.1	3	0.05	0.5
Bending strength	N/mm^2	15	30	–	100	–	15
Tensile strength	N/mm^2	12	15	20	50	15...30	20
Thermal conductivity λ	W/K m	0.3	0.4	0.3	0.17	0.17	0.25
Linear thermal expansion	10^{-6}/K	320	150	320	70	190	120
Thermal stability limit	°C	70	80	90	65	50	250

Table A 3.6: Properties of cast-resin mouldings

Property	Unit	EP-mouldings FS 1000-0	FS 1000-6	FS 1021-0	FS 1021-6	PUR
Resin component		EP-solid resin	EP-solid resin	EP-liquid resin	EP-liquid resin	–
Hardener[1])		PSA	PSA	HH PSA	HH PSA	–
Filler	% SiO_2	0	60	0	65	–
Hardening temperature	°C	120	120	60/90	60/90	–
Breakdown field strength E_d	kV/mm	15	15	15	16	15
Volume resistivity ρ	Ωcm	10^{14}	10^{14}	10^{14}	10^{14}	10^{13}
Dielectric constant ϵ_r		3	4	4	4	4
Dissipation factor (1 MHz) tan δ		10^{-2}	10^{-2}	10^{-2}	10^{-2}	$2 \cdot 10^{-2}$
Tracking index		KA 3a	KA 1	KA 3c	KA 3c	KA 3c
Density γ	g/cm^3	1.2	1.8	1.2	1.9	1.2
E-modulus	kN/mm^2	4	12	2.5	12	0.5
Bending strength	N/mm^2	140	135	120	110	–
Tensile strength	N/mm^2	80	90	60	75	3
Thermal conductivity λ	W/Km	0.2	0.8	0.2	0.8	0.24
Linear thermal expansion	10^{-6}/K	65	35	65	35	200
Thermal stability limit	°C	125	130	130	130	80
Arc withstand index		L1	L1	L4	L1	–

[1]) PSA: Phthalic acid anhydride; HHPSA: Hexahydrophthalic acid anhydride
According to DIN 16946, Part 2, the following designations mean:
FS 1000-0: unfilled moulding of solid EP
FS 1000-6: 60% filled moulding of solid EP
FS 1021 : liquid EP

Appendix 4

Properties of magnetic materials

The essential parameters for dimensioning equipment with iron cores are the magnetic properties of the core materials. The non-linear relationship between the flux density $\hat{B}$ and the field strength H can be described by the magnetization characteristic $\hat{B} = f(H)$ or by the permeability $\mu_r = f(\hat{B})$. We have:

$$\hat{B} = 2\mu_0\mu_r H\sqrt{2} \quad \text{with} \quad \mu_0 = 0.4\pi \cdot 10^{-6} \frac{Vs}{Am}.$$

The losses in the iron core can be represented by the iron loss angle $\delta_E = f(\hat{B})$ or losses referred to unit mass $P'_{Fe} = f(\hat{B})$. Both these quantities are related in the magnetization characteristic according to the following equation:

$$\sin\delta_E = \frac{P'_{Fe}(\hat{B})\gamma}{H(\hat{B})\omega\,\hat{B}/\sqrt{2}} \qquad \begin{array}{l}\gamma = \text{density}\\ \omega = \text{angular frequency}\end{array}$$

These characteristics are measured for appropriate samples by the manufacturer and published. A typical example was chosen from this data for a grain-oriented silicon-iron alloy (1, TRAFOPERM N2) and a nickel-iron alloy (2, MUMETALL) [*Vakuumschmelze* 1977].

The cores can be made of wound tape (wound ring core) or built of laminated sheets. Wound ring cores have different values for the characterising parameters than laminated cores. The curves shown are valid for 50 Hz alternating voltage and sheet thicknesses between 0.2...0.35 mm.

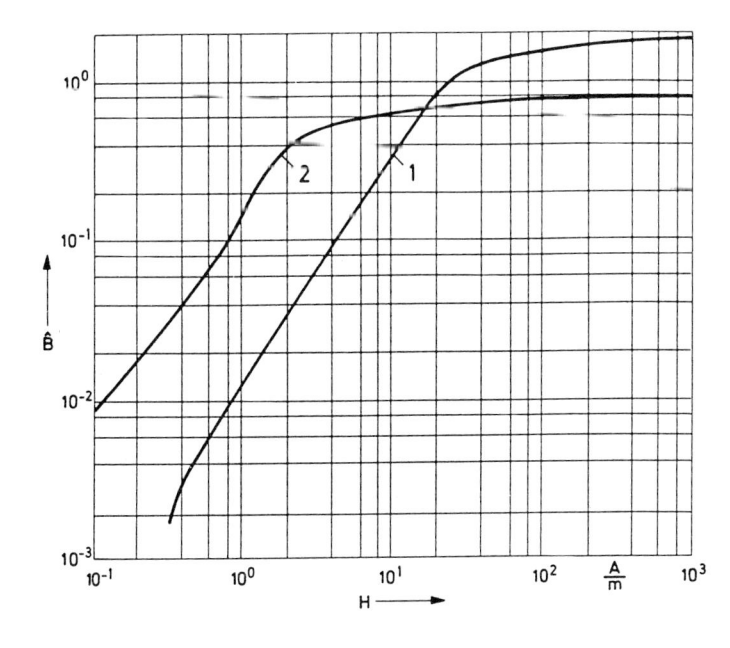

Fig. A 4-1

Magnetization curve for wound ring cores

178   Appendix 4

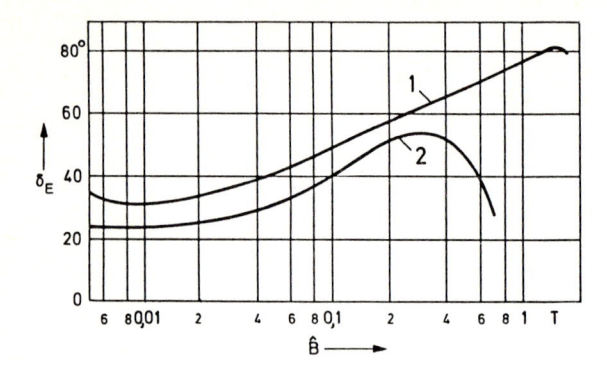

Fig. A 4-2
Iron loss angle of wound ring cores

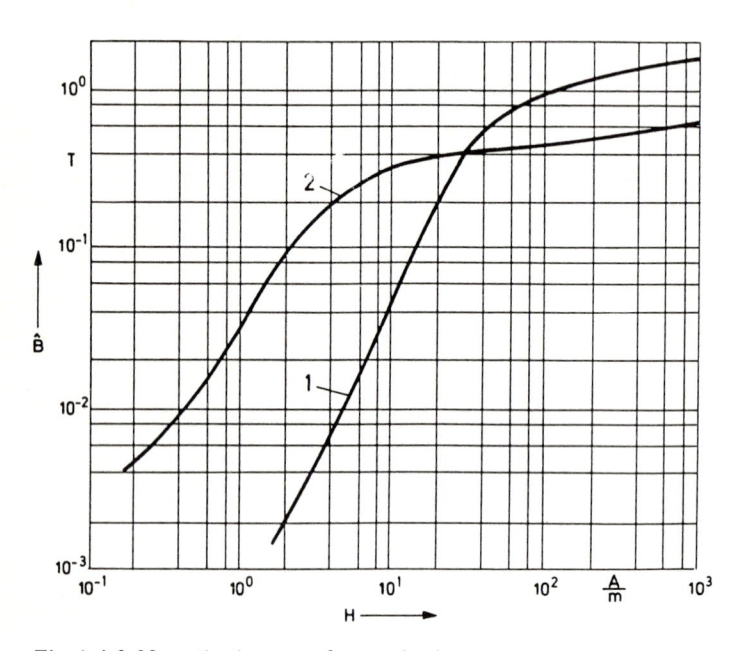

Fig. A 4-3 Magnetization curve for core laminae

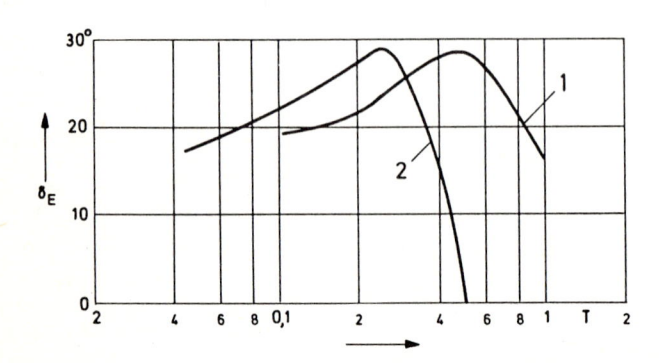

Fig. A 4-4
Iron loss angle of core laminae

Properties of magnetic materials

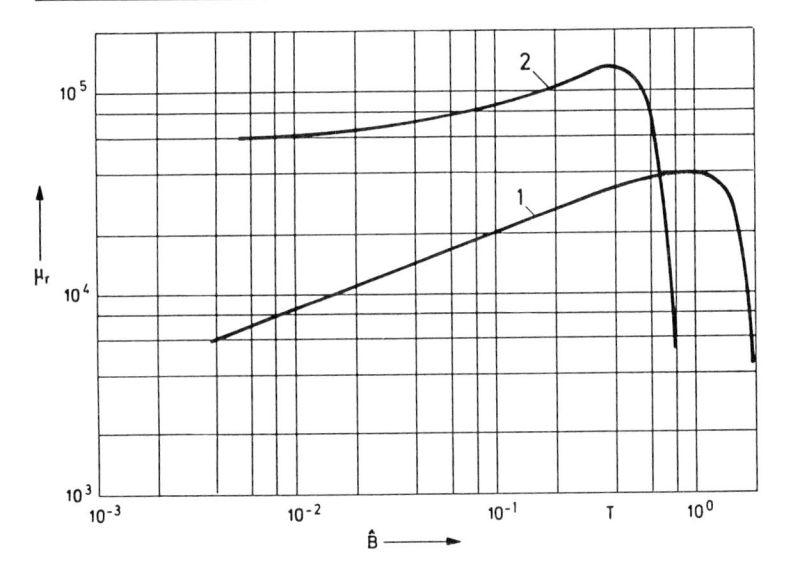

Fig. A 4-5 Permeability of wound ring cores

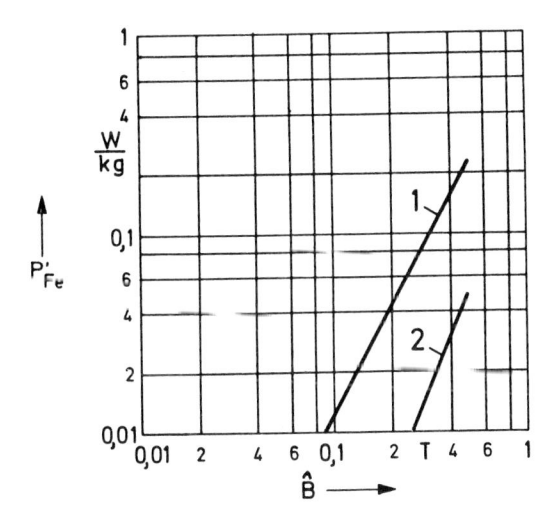

Fig. A 4-6
Specific magnetization losses of core laminae

Bibliography

Abou-Seada, M. S., Nasser, E., Digital computer calculation of the potential and field of a rod gap. IEEE Proc. 56 (1968), p. 813–820.

Alpert, D., Lee, D. A., Lyman, E. M., Tomaschke, H. W., Initiation of electrical breakdown in ultrahigh vacuum. J. Vac. Sci. Technol. Vol. 1 (1964), p. 35–50.

Alston, L. L., High voltage technology. Oxford Univ. Press 1968.

Ambrozie, C., Berechnung der Ersatzkapazitäten der verschachtelten Scheibenspulentransformatorwicklungen. E. u. M. 92 (1975), 1, p. 23–35.

Andersen, O. W., Laplacian electrostatic field calculations by finite elements with automatic grid generation. IEEE Trans. PAS 92 (1973), p. 1485–1492.

Anderson, J. C., Dielectrics. Chapman and Hall, London 1964.

Ari, N., Strahlungsfelder in Hochspannungs-Meßanordnungen für Stoßspannungen. Bull. SEV 65 (1974), No. 16, p. 1222–1229.

Artbauer, J., Elektrische Festigkeit von Polymeren. Kolloid-Z. und Z. für Polymere 202 (1965), 1, p. 15–26.

Artbauer, J., Die elektrische Festigkeit von Polymeren als Extremgröße. Kolloid-Z. und Z. für Polymere, Bd. 225 (1968), 1, p. 23–29.

Baer, G., Lehmann, W., SF_6 insulated metalclad 420 kV tubular bus with transfer facility connected to an underground hydroelectric power station. CIGRE-Ber. 1974, No. 23–02.

Bahder, G., Dakin, T. W., Lawson, J. H., Analysis of treeing type breakdwon. CICRE-Ber. 1974, No. 15–05.

Bartnikas, R., McMahon, E. J., Engineering dielectric Vol.I, Corona measurement and interpretation. American Society of Testing and Materials, Philadelphia, USA, 1979.

Bauer, E., Der Hochvakuumdurchbruch zwischen Wolframelektroden bei hoher Stoßspannung. Dissertation TU München 1971.

Bauer, R., Die Meßwandler. Springer-Verlag, Berlin 1953.

BBC, BBC Taschenbuch für Schaltanlagen. 6. Auflage, Verlag W. Girardet, Essen 1977.

Berger, K., Zur Theorie des Wärmegleichgewichts fester Isolatoren. ETZ 47 (1926), p. 673–677.

Berger, K,, Der Durchschlag fester Isolierstoffe als Folge ihrer Erwärmung. Bull. SEV 17 (1926), No. 2, p. 37–57.

Beyer, M., Grundlagen der Vakuumtrocknung von flüssigen Isolierstoffen und Papierisolation. VDI-Bildungswerk BW 870, 1971.

Biermanns, J., Hochspannung und Hochleistung. Carl Hanser, München 1949.

Binns, K. J., Lawrenson, P. J., Analysis and computation of electric and magnetic field problems. Pergamon Press, Oxford 1963.

Boag, J. W., The design of the electric field in a van de Graaff generator. Proc. IEE IV 100 (1953), 5, p. 63–82.

Boeck, W., Entstehung und Bedeutung von Raumladungen in Kunststoff-Folien durch Koronaentladungen. Dissertation TU Braunschweig 1967.

Böcker, H., Reichert, K., Digitale Berechnung von elektrischen Feldern in metallgekapselten Anlagen. ETZ-A 94 (1973), p. 374–377.

Böning, P., Kleines Lehrbuch der elektrischen Festigkeit. Braun, Karlsruhe 1955.

Böttger, O., Langzeitdurchschlagsverhalten von Polyäthylen an Kabeln. Bull. SEV 64 (1973), No. 3, p. 143–148.

Bogorodizki, N. P., Pasynkow, W. W., Tarejew, B. M., Werkstoffe der Elektrotechnik. VEB-Verlag, Berlin 1955.

Brand, U., Kind, D., Gas impregnated plastic foils for high voltage insulation. CIGRE-Ber. 1972, No. 15–02.

Bibliography

Brand, U., Hochspannungsisolierungen mit gasimprägnierten Kunststoff-Folien. Dissertation TU Braunschweig 1973.
Brechna, H., Stoßspannungssichere Transformatorwicklungen. Bull. Oerlikon, No. 328/329, August 1958, p. 89–100.
Brinkmann, C., Die Isolierstoffe der Elektrotechnik. Springer-Verlag, Berlin 1975.
Büsch, W., Der Einfluß der Luftfeuchtigkeit auf die Schaltspannungsfestigkeit der atmosphärischen Luft im Ultrahochspannungsbereich bei positivier Polarität. Diss. ETH-Zürick 1982.
Caldwell, R. O., Darveniza, M., Experimental and analytical studies of the effect of non-standard waveshapes on the impulse strength of external insulation. IEEE PAS 92 (1973), No. 4, p. 1420–1428.
Carrara, G., Dellera, L., Accuracy of an extended up-and-down method in statistical testing of insulation. Electra 1972, No. 23, p. 159–179.
Cranberg, L., The initiation of electrical breakdown in vacuum. J. Appl. Phys., Vol. 23 (1952), p. 518–522.
Dakin, T. W., Luxa, G., Oppermann, G., Vigreux, J., Wind, G., Winkelnkämper, H., Breakdown of gases in uniform fields. Electra 1974, No. 32, p. 61–82.
Densley, R. J., Bulinksi, A., Robert, J., The surge characteristics of XLPE insulation containing water trees. Conference Record of the 1980 IEEE International Symposium on Electrical Insulation.
Dieterle, W., Schirr, J., Elektrische und mechanische Eigenschaften von im Druckgelierverfahren hergestellter Epoxidharzformstoffe. Bull. SEV Bd. 63 (1972), No. 22, p. 1300–1304.
Dittmer, B., Der räumliche und zeitliche Entladungsaufbau in festen Isolierstoffen im ungleichförmigen Feld. Archiv für Elektrotechnik 48 (1963), part 1: p. 150–166, part 2: p. 287–296, part 3: p. 387–402.
Dohnal, D., Untersuchungen zur Röntgenstrahlung an Hochspannungs-Hochvakuum-Anordnungen. Dissertation TU Braunschweig 1981.
Dokopoulos, P., Die Durchschlagswahrscheinlichkeit von Hochspannungsisolierungen. ETZ-A 89 (1968), 7, p. 145–150, s. auch Diss. TU Braunschweig 1967.
Dokopoulos, P., Steudle, W., Über die Anwendung von Wasser zur Isolierung von Hochspannungs-Impulsanlagen. ISH München 1972, p. 599–606.
Dreyfus, L., Mathematische Theorien für den Durchschlag fester Isoliermaterialien. Bull. SEV 13 (1924), No. 7, p. 123–145.
Dronsek, G., Nebendurchschläge in festen Isolierstoffen bei Wechselspannung. Dissertation TU Braunschweig 1967.
Dyke, W. P., Dolan, W. W., Trolan, J. K., The field emission initiated vacuum arc I. Phys. Rev. Vol. 91 (1953), p. 1043.
Ehlers, W., Lau, H., Kabelherstellung. Springer-Verlag, Berlin 1956.
Erk, A , Schmelzle, M., Grundlagen der Schaltgerätetechnik Springer-Verlag, Berlin 1974.
Feser, K., Bemessung von Elektroden im UHV-Bereich, gezeigt am Beispiel von Toroidelektroden für Spannungsteiler. ETZ-A Bd. 96 (1975), 4.
FHG Forschungsgemeinschaft für Hochspannungs- und Hochstromtechnik e.V., Die elektrische Festigkeit von Freiluftisolationen bei Blitzüberspannungen. Technischer Bericht 1–243, 1979.
Finkelnburg, W., Einführung in die Atomphysik. Springer-Verlag, Berlin 1967.
Fock, V., Zur Wärmetheorie des elektrischen Durchschlags. Archiv für Elektrotechnik 19 (1927), p. 71–81.
Franz, W., Dielektrischer Durchschlag. Handbuch für Physik von S. Flügge, Bd. 17, Springer, Berlin 1956.
Gänger, B., Der elektrische Durchschlag in Gasen. Springer-Verlag, Berlin 1953.
Gänger, B., Der Flüssigkeitsdurchschlag. Bull. SEV 72 (1981), No. 13, p. 680–689.
Galloway, R. H., McRyan, H., Scott, M. F., Calculation of electric fields by digital computer. Proc. IEE 114 (1967), p. 824–829.
Georg, G., Kunststoff-Filme für Leistungskondensatoren. Bull. SEV 60 (1969), No. 14, p. 630–634.
Gerhold, J., Überlegungen zur elektrischen Isolation supraleitender Wellrohrkabel. ETZ-A 98 (1977), 10, p. 685.
Gertsch, R., Transformateurs de tension capacitifs et leur fonctionnement avec les relais de protection de rèseau. CIGRE-Ber. 1960, No. 318.

Guthmann, R., Die Spannungsabhängigkeit des Verlustfaktors bei Folien-Papierkondensatoren. ETZ-A 75 (1954), 2, p. 45–48.

Halleck, M. C., Calculation of corona-starting voltage in air-solid dielectric systems. AIEE Trans., Vol. 7 (1956), OAS 23–28, p. 211–216.

Hanella, K., Beitrag zur Ermittlung des Durchschlagsverhaltens dünner Folien. Wiss. Z. der Elektrotechnik 10 (1968), p. 209–228.

Happoldt, H., Oeding, D., Elektrische Kraftwerke und Netze. 5. Auflage, Springer-Verlag, Berlin 1978.

Heinhold, L., Kabel und Leitungen für Starkstom. Siemens AG 1965.

Hermstein, W., Entwicklungstendenzen im Wandlerbau. Elektrizitätswirtschaft 68 (1969), 8, p. 246–257.

Hersping, A., Über einige Eigenschaften von Kunststoffen für die Elektrotechnik. Kunststoffe 52 (1962), 2, p. 73–77

Hess, H., Der elektrische Durchschlag in Gasen. Vieweg, Braunschweig 1976.

Heumann, H., Patsch, R., Saure, M., Wagner, H., Observations on water-treeing especially at interfaces of polyolefine cable insulation. CIGRE-Ber. 1980, No. 15–06.

v. Hippel, A., Dielectrics and waves. Wiley, New York 1954.

v. Hippel, A., Dielectric materials and applications. 2. edition, Wiley, New York 1958.

Holle, K.-H., Über die elektrischen Eigenschaften von Isolierölen, insbesondere über den Einfluß von Wasser auf deren Temperaturverhalten. Dissertation TH Braunschweig 1967.

Holte, K. C. u.a., Application of insulators in a contaminated environment. IEEE PAS 98 (1979), No. 5, p. 1676–1695.

Holzmüller, W., Altenburg, K., Physik der Kunststoffe. Akademie-Verlag, Berlin 1961.

Hütte,, Hütte, Des Ingenieurs Taschenbuch, Bd.IVA, Starkstromtechnik, Lichttechnik. Wilhelm Ernst & Sohn, Berlin 1957.

Hyltén-Cavallius, N., Chagas, F. A., Possible Precision of Statistical Insulation Test Methods. IEEE PAS-102 (1983), No. 8, p. 2372–2378.

Ieda, M., Sawa, G., Miyairi, K., Dielectric breakdown of polyethylene films at cryogenic temperature. ISH München 1972, p. 414–420.

Imhof, A., Hochspannungs-Isolierstoffe. Braun, Karlsruhe 1957.

Jähne, H., Über den Einfluß innerer mechanischer Spannungen auf die Leitungsmechanismen in Epoxidharz bei Gleich- und Wechselspannung. Dissertation TU Hannover 1975.

Kappeler, H., Hartpapierdurchführungen für Höchstspannung. Bull. SEV 40 (1949), No. 21, p. 807–815.

Kappeler, H., Resin-bonded paper bushings for EHV-systems. IEEE Trans. PAS 88 (1968), p. 394–399.

Kind, D., Die Aufbaufläche bei Stoßbeanspruchung technischer Elektrodenanordnungen in Luft. Dissertation TH München 1957, s. auch ETZ-A 79 (1958), 3, p. 65–69.

Kind, D., Hermetischer Luftabschluß ölisolierter Hochspannungsgeräte mittels Stickstoffpolster. Elektrizitätswirtschaft 58 (1959), 5, p. 143–149.

Kind, D., Die Hochspannungstechnik am Beginn des Kunststoffzeitalters. ETZ-A 91 (1971), p. 134–139.

Kind, D., Einführung in die Hochspannungs-Versuchstechnik. Vieweg, Braunschweig 1972 (2. Auflage 1978).

Knörrich, K., Koller, A., Digitale Berechnung von ebenen und rotationssymmetrischen Potentialfeldern mit beliebigen Randbedingungen. ETZ-A 91 (1970), 6, p. 339–344.

Knorr, W., Der Stoßdurchschlag koaxialer Elektrodenanordnungen in SF_6 bei kurzen Durchschlagszeiten. ETZ-A 98 (1977), p. 549–551.

Knust, E., Chlorierte Dielektrika als Tränkmittel für Papierkondensatoren. E u. M 82 (1965), 3, p. 132–134.

König, D., Erfassung von Teilentladungen in Hohlräumen von Epoxidharzplatten zur Beurteilung des Alterungsverhaltens bei Wechselspannung. Dissertation TU Hannover 1967.

König, D., Lipken, H., Weltgen, J., Untersuchungen an Freiluft-Durchführungen mit Feststoff-Hauptisolation für metallgekapselte SF_6-isolierte Schaltanlagen bis 150 kV Betriebsspannung. ETZ-A 98 (1977), 11, p. 761.

Bibliography

Koermann, W., Über den elektrischen Durchschlag in Luft bei kleinen und mittleren Überspannungen. Zeitschrift für angewandte Physik 7 (1955), 4, p. 187–194.

Kodoll, W., Teilentladungs-Durchschlag von polymeren Isolierstoffen bei Wechselspannung. Dissertation TU Braunschweig 1974.

Kok, J. A., Der elektrische Durchschlag in flüssigen Isolierstoffen. Philips Techn. Bibliothek, Eindhoven 1963.

Kolossa, I., Zusammenhänge zwischen den für Fremdschichtüberschläge maßgebenden Kenngrößen von Hochspannungsanlagen. ATM (1971), 12, p. 247–252 und ATM (1972), 1, p. 15–18.

Kratzenstein, M. G., Der Stoßdurchschlag in Isolieröl. Dissertation TH München 1969, s. auch Bull. SEV 61 (1970), 3, p. 105–117.

Kreuger, F. H., Discharge detection in high voltage equipment. American Elsevier Publ. Co., New York 1964.

Kubenz, R., Martin, H. D., Kunststoffe für hohe elektrische Beanspruchungen. Kontakt und Studium Bd. 6, Lexika-Verlag, Grafenau 1976.

Kübler, B., Teilentladungsverhalten von Modellanordnungen aus Epoxidharz-Formstoff mit natürlichen Fehlstellen. Dissertation TU Braunschweig 1978.

Küchler, R., Die Transformatoren. Springer, Berlin 1966.

Kuffel, E., Zaengl, W. S., High-Voltage Engineering. Pergamon Press, 1984.

Kuzhekin, I. P., Verhalten von Funkenstrecken unter Wasser bei Impulsspannung und Impulsstrombeanspruchung. EZT-A 93 (1972), p. 404–409.

Lambeth, P. J., Effect of pollution on high-voltage outdoor insulators. IEE Review, Vol. 118 (1971), p. 1107–1130.

Latham, R. V., High voltage vacuum insulation: The physical basis. Academic Press, London 1981.

Lautz, G., Elektromagnetische Felder. Teubner, Stuttgart 1969 (2. Auflage 1976).

Lemke, E., Ein Beitrag zur Abschätzung der räumlich-zeitlichen Entwicklung des Durchschlags in langen Luftfunkenstrecken bei Schaltspannungen. Wiss. Zeitschrift der TU Dresden 26 (1977), 1, p. 133–136.

Lesch, G., Lehrbuch der Hochspannungstechnik. Springer, Berlin 1959.

van Leyen, Wärmeübertragung. Siemens AG, Berlin, München 1971.

Liebscher, F., Held, W., Kondensatoren. Springer, Berlin 1968.

Link, W. D., Überschlag von Stützisolatoren in Luft in Abhängigkeit von Temperatur und Luftfeuchte. Dissertation Universität Stuttgart 1975.

Llewellyn-Jones, F., Ionization and breakdown in gases. Science Paperbacks, London 1957.

Loeb, L. B., Basic processes of gaseous electronics. University of California Press, Los Angeles and Berkely 1960.

Löffelmacher, G., Über die physikalisch-chemischen Vorgänge bei der Ausbildung von Endladungskanälen in Polyäthylen und Epoxidharz im inhomogenen Wechselfeld. Dissertation TU Hannover 1976.

Lücking, W. W., Energiekabeltechnik. Vieweg, Braunschweig 1981.

Lührmann, H., Rasch veränderliche Vorgänge in räumlich ausgedehnten Hochspannungskreisen. Dissertation TU Braunschweig 1973, s. auch Archiv für Elektrotechnik 57 (1975), p. 253–264.

Lutz, H., Transformer oil preservation systems and associated problems. CIGRE-Ber. 1960, No. 134.

Luy, H., Oswald, F., Die Durchschlagsfestigkeit von Polyäthylen. ETZ-A 92 (1971), 6, p. 358–363.

Maitland, A., Influence of the anode temperature in the breakdown voltage and conditioning characteristic of a vacuum gap. Brit. J. Appl. Phys. Vol. 13 (1962), p. 122–125.

Meek, J. M., Craggs, J. D., Electrical breakdown of gases. Clarendon Press, Oxford 1978.

Menges, G., Berg, H., Durchschlagsmechanismus und Verformungszustände. ETZ-B 24 (1972), 25, p. 643–645.

Mierdel, G., Der elektrische Durchschlag. Elektrie 21 (1967), 12, p. 483–487.

Moeller, J., Steinbigler, H., Weiß, P., Feldstärkeverlauf auf Abschirmelektroden für ultrahohe Wechselspannungen. ISH München 1972.

Morva, T., Verfahren zum Berechnen der elektrischen Feldstärke an Hochspannungselektroden. ETZ-A 87 (1966), 26, p. 955–959.

Mosch, W., Pilling, J., Tschacher, B., Eckholz, K., Kanaleinsatzzeit und Durchschlagszeit zur Beurteilung des Langzeitverhaltens von Feststoffisolierungen. Elektrie 26 (1972), 11, p. 319–322, s. auch Wiss. Beiträge ISH München 1972, p. 421–426.

Mosch, W., Hauschild, W., Hochspannungsisolierungen mit Schwefelhexafluorid. Dr. A. Hüthig-Verlag, Heidelberg, Basel 1978.

Moser, H. P., Transformerboard. Birkhäuser AG, Basel 1979.

Müller, R., Molitov, W., Eigenschaftsänderung der Isolierstoffe von Öltransformatoren unter betrieblichen Beanspruchungen. ETG-Fachberichte Dauerverhalten von Isolierstoffen und Isoliersysteme. VDE-Verlag, Berlin 1977.

Müller, W., Unkonventionelle Meßwandler für Höchstspannungen. ETZ-A 93 (1972), 6, p. 362–366.

Müller, W., Die räumliche und zeitliche Stoßspannungsverteilung in Transformatorwicklungen. Bull. SEV 66 (1975), No. 4, p. 218–224, s. auch Diss. TU Braunschweig 1972.

Naglik, M., Niederinduktive kapazitive Energiespeicher mit hohen Spannungen. Dissertation TU Braunschweig 1973.

Nasser, E., Zum Problem des Fremdschichtüberschlages an Isolatoren. ETZ-A 83 (1962), 11, p. 356–365.

Nasser, E., Fundamentals of gaseous ionization and plasma electronics. Wiley-Interscience 1971.

Nitta, T., Shibuya, Y., Electrical breakdown of long gaps in sulfurhexafluoride. Trans. IEEE, Vol. PAS 90 (1971), p. 1065–1069.

Oburger, W., Die Isolierstoffe der Elektrotechnik. Springer-Verlag, Wien 1957.

Patsch, R., Wagner, H., Heumann, H., Inhomogeneities and their significance in single-layer extruded polyolefine insulations for cables. CIGRE-Ber. 1976, No. 15–11.

Peier, D., Die Einleitung des elektrischen Durchschlags in verflüssigten, tiefsiedenden Gasen. Archiv für Elektrotechnik 58 (1976), p. 39–46.

Peschke, E., Hochleistungsübertragung mit Kabeln. Siemens Forschungs- und Entwicklungsberichte Bd. 2 (1973), 1.

Peschke, E., Einleiter-Ölkabel und Garnituren für 380 kV. Siemens-Zeitschrift 50 (1976), 10, p. 676–684.

Philippow, E., Taschenbuch Elektrotechnik, Bd. 2 (Starkstromtechnik). VEB-Verlag, Berlin 1966.

Philippow, E., Taschenbuch Elektrotechnik, Bd. 1 (Allgemeine Grundlagen). VEB-Verlag, Berlin 1976.

Prinz, H., Singer, H., Der Schwaigersche Ausnutzungsfaktor grundlegender translatorischer und rotatorischer Elektrodensysteme. Bull. SEV 58 (1967), No. 4, p. 161–167.

Prinz, H., Hochspannungsfelder. Oldenbourg, München 1969.

Raether, H., Electron avalanches and breakdown in gases. Butterworths, London 1964.

Ragaller, K., Surges in high-voltage networks. Plenum Press, New York 1980.

Rao, Y. N., Die Alterung von gasimprägnierten Folienisolierungen durch Koronaentladungen bei Wechselspannungen. Dissertation TU Braunschweig 1968.

Rayes, M. N., Untersuchungen an Überschlagsanordnungen in Luft mit Teilstrecken von unterschiedlichem Entladungsverhalten bei Wechselspannung. Dissertation TU Braunschweig 1978.

Reichert, K., Über Verfahren zur numerischen Berechnung elektrostatischer Felder. ETZ-A 93 (1972), 6, p. 338–339.

Rein, A., Breakdown mechanisms and breakdown criteria in gases. Electra 1974, No. 32, p. 43–60.

Renardières-Group, Research on long air gap discharges at Les Renardières – 1973 results –. Electra 1974, No. 35, p. 49.

Renardièries-Group, Positive discharges in long air gaps at Les Renardières – 1975 results and conclusions –. Electra 1977, No. 53, p. 31.

Reverey, G., Verma, M. P., Fremdschicht-Prüfverfahren und Untersuchungen an verschmutzten Isolatoren im In- und Ausland. ETZ-A 91 (1970), 9, p. 481–488.

Richter, R., Elektrische Maschinen, Bd. III, Transformatoren. Birkhäuser Verlag, Basel 1963.

Richter, R., Elektrische Maschinen, Bd. I, Allgemeine Berechnungselemente. Gleichstrommaschine. Birkäuser Verlag, Basel 1967.

Rosenberger, G., Betriebseigenschaften kapazitiver Spannungswandler. ETZ-A 87 (1966), 15, p. 556–560.

Bibliography

Roth, A., Hochspannungstechnik 5. Auflage. Springer-Verlag, Berlin 1965.
Ryan, H. M., Mattingley, J. M., Scott, M. F., Computation of electric field distributions in high-voltage equipment. IEEE trans. on El. Insul. EI-6 (1971), No. 4, p. 148–154.
Saechtling, H., Zebrowski, W., Kunststoff-Taschenbuch, 18. Auflage. Carl Hanser Verlag, München 1971.
Salvage, B., Hiley, J., El-Gendy, O. A., Stuttock, I. R., McGuinness, A. M., A study of the effects of internal discharges on an epoxy resin using a scanning electron microscope. Internationales Symposium Hochspannungstechnik, Zürich 1975.
Saure, M., Kunststoffe in der Elektrotechnik. AEG-Telefunken Handbuch 1979.
Schirr, J., Beeinflussung der Durchschlagsfestigkeit von Epoxidharz-Formstoff durch das Herstellungsverfahren und durch mechanische Spannungen. Dissertation TU Braunschweig 1974.
Schiweck, L., Über den Einfluß von Elektrodenanordnung und Spannungsverlauf auf den Durchschlag von Epoxidharz-Formstoff. Dissertation TU Braunschweig 1969.
Schlosser, K., Eine auf physikalischen Grundlagen ermittelte Ersatzschaltung für Transformatoren mit mehreren Wicklungen. BBC-Nachrichten 1963, p. 107–132.
Schmidt, B.-D., Untersuchungen zum Durchschlag von Hochvakuumanordnungen bei kryogenen Temperaturen. Dissertation TU Braunschweig 1979.
Schneider, K.-H., Weck, K.-H., Parameters influencing the gap factor. Electra No. 35 (1974), p. 25–45.
Schon, K., Salpetersäureaufnahme von Kunststoff-Folien unter Einwirkung von Teilentladungen. ETZ-A 98 (1977), p. 504–506.
Schühlein, E., Zur Wanddickenabhängigkeit der Durchschlagsspannung von Epoxidharz-Formstoffisolierungen. ETZ-B 20 (1968), 3, p. 363–367.
Schumann, W. O., Elektrische Durchbruchsfeldstärke von Gasen. Theoretische Grundlagen und Anwendungen. Springer-Verlag, Berlin 1923.
Schwaiger, A., Elektrische Festigkeitslehre. Springer-Verlag, Berlin 1925.
Shihab, S., Teilentladungen in Hohlräumen von polymeren Isolierstoffen bei hoher Gleichspannung. Dissertation TU Braunschweig 1972.
Shortley, G., Weller, R., Darby, P., Gamble, E. H., Numerical solution of axisymmetrical problems with applications to electrostatics and torsion. J. Appl. Phys. 18 (1947), p. 116–129.
Singer, H., Das elektrische Feld von Gitterelektroden. ETZ-A 90 (1969), 25, p. 682–686.
Singer, H., Flächenladungen zur Feldberechnung von Hochspannungssystemen. Bull. SEV 65 (1974), p. 739–746.
Singer, H., Steinbigler, H., Weiß, P., A charge simulation method for the calculation of high voltage fields. IEEE Trans. PAS 93 (1974), p. 1660–1668.
Sirotinski, L. I., Hochspannungstechnik, Bd. I, Teil 1, Gasentladungen. VEB-Verlag, Berlin 1955.
Sirotinski, L. I., Hochspannungstechnik, Bd. II, Isolatoren und Isolierungen. VEB-Verlag, Berlin 1958.
Southwell, R. V., Relaxation methods in engineering science. Oxford Univ. Press 1949.
Specht, H., Untersuchungen zum Einfluß von Isolierstoffoberflächen auf den Durchschlag in Schwefelhexafluorid. Dissertation TU Braunschweig 1977.
Steinbigler, H., Digitale Berechnung elektrischer Felder. ETZ-A 90 (1969), 25, p. 663–666.
Steudle, W., Über die Anwendung von Wasser zur Isolierung von Hochspannungsimpulsanlagen. Dissertation TU Braunschweig 1974.
Strigel, R., Winkelnkemper, H., Die Durchschlagsfestigkeit von luftübersättigtem Öl und Öl-Papier-Dielektrikum. ETZ-A 82 (1961), 26, p. 833–838.
Strigel, R., Elektrische Stoßfestigkeit. Springer, Berlin 1955.
Taegen, F., Einführung in die Theorie der elektrischen Maschinen I. Vieweg, Braunschweig, 1970.
Thoma, P., Instabilities during high-field electrical conduction in solids. J. Appl. Phys. 47 (1976), No. 12, p. 5304–5312.
Tiedemann, W., Werkstoffe für die Elektrotechnik. Band II, Nichtmetallische Werkstoffe. VEB-Verlag Technik, Berlin 1962.
Unger, H. G., Schultz, W., Elektronische Bauelemente und Netzwerke I. Vieweg, Braunschweig 1968.
Vakuumschmelze, Handbuch Weichmagnetische Werkstoffe. Vakuumschmelze GmbH, Hanau 1977.
Vajda, G., Acta Technica Academicae. Scientiarium Hungaricae 56 (1966), 3–4, p. 319–332.

VDEW, VDEW-Kabelhandbuch. Verlags- und Wirtschaftsgesellschaft der Elektrizitätswerke mbH, Frankfurt 1977.

VDI, VDI Wärmeatlas, Berechnungsblätter für den Wärmeübergang. VDI Verlag GmbH, Düsseldorf 1974.

Verma, M. P., Die quantitative Erfassung von Fremdschichteinflüsse. ETZ-A 97 (1976), 5, p. 281–285.

Vogelmann, M., Henny, J. D., Langzeitverhalten von Epoxidgießharzen. Symposium Elektrische Isolationstechnik, Band I, Zürich 1972.

Wagner, H., Zum elektrischen Durchschlag von teilkristallinen Polymeren. ETZ-A 94 (1973), 7, p. 436–437.

Weck, K. H., Fischer, A., Dielectric strength of insulations under non-standard lightning impulses. CIGRE 33–75 (SC) 04 IWD, Colloquium Montreal 1975, Canada.

Weiss, M., Finite Elemente – Ein Verfahren zur Lösung komplizierter Feldprobleme. ETZ-A 95 (1974), 9, p. 462–463.

Weiß, P., Berechnung von Zweistoffdielektrika. ETZ-A 90 (1969), p. 693–694.

Weniger, M., Über die Zündung von Teilentladungen in polymeren Isolierstoffen durch Wechsel- und Stoßspannungen. Dissertation TU Braunschweig 1975.

Whitehead, J. B., Impregnated paper insulation. Wiley & Sons, New York 1935.

Whitehead, S., Dielectric breakdown of solids. Clarendon Press, Oxford 1951.

Widmann, W., Das Vergrößerungsgesetz in der Hochspannungstechnik. ETZ-A 85 (1964) 4, p. 97–102.

Wilhelmy, L., Eine Sonde zur potentialfreien Messung der periodischen und transienten Feldstärke. ETZ-A 94 (1973), 8, p. 441–445.

Zaengl, W., Nyffenegger, H., Berechnung der Koronaeinsetzfeldstärke zylindrischer Leiter in Luft. Bull. SEV 65 (1974), No. 12, p. 873–879.

Zeibig, A., Der Hochvakuumdurchbruch bei hoher Gleichspannung. Dissertation TH München 1966.

Zienkiewicz, O. C., Numerical solution of 3-dimensional field problems. Proc. IEE 115 (1968), p. 367–369.

Zinke, O., Widerstände, Kondensatoren, Spulen und ihre Werkstoffe. Springer-Verlag, Berlin 1965.

Zinn, E., PTB-Prüfregeln, Band 12, Meßwandler. PTB, Braunschweig 1977.

Zoledziowski, S., Soar, S., Life curves of epoxy resin under impulses and the breakdown parameters. IEEE Trans. El. Insul. EI-7 (1972), No. 2, p. 84–99.

Index

absorption of moisture 70
a.c. capacitors 115
adsorption isotherms 83
ageing of insulating materials 45, 47, 79
– products 30, 80
air density correction 21, 169, 171
– sealing 108
alumina 78
– porcelain 77
angle of contact 70
– -rings 138, 141
anomalous breakdown 47
arc extinction 47, 73
arcing resistance 67
askarels 94
attachment coefficient 18

band model (of an insulator) 31
bellows 109
boundary surface 98
breakdown 8, 14
– field strength 9, 19, 63
– – – of gas-insulated configurations 168
– – – insulating materials 173
– – – transformer oil 48
–, high-temperature 33
–, high vacuum 52
– in a strongly inhomogeneous field 22, 35
– of gases 14, 168
– – liquid insulating materials 47
– – solid insulating materials 30
– probability 10
–, streamer mechanism 20
– strength 9
– time 27
– under lightning impulse voltages 25
– – switching impulse voltages 28
– voltage 1, 8, 11, 168, 171
– – distribution 11, 24
brush discharges 22, 29, 124
burden impedance (load impedance) 144
bushings 100, 120

cable insulation 84
– paper 83
– termination 122
cam-mounting 107
capacitive control 105, 121
– voltage transformer 148
capacitor paper 81

capacitors 111
cap insulators 74, 101
carrier current 16
cascade connection 135, 148
casting plant 92
cast resins 91
–, properties 176
cemented flange 107
charge carriers
– in gases 14
– in solid insulating materials 30
–, injection of 45
–, production of 18
–, simulation method 3, 5
chlorinated diphenyls 94
circular ring-configurations 165
coaxial support insulators 102
coil winding 135, 140
collision ionization 16, 22, 32, 50
compound dielectric 82
– -filled cable 84
conductive coating 120, 122
connecting tube 101
contamination layer 54, 71, 98, 101
– model 58
continuous corona 24
convection 110
corona discharges 9, 24
coupling 105
– capacitors 118, 151
creepage path 57, 100
critical voltage of thermal breakdown 38
current error 154
– transformer 152
cylindrical configurations 161
– windings 113, 135

d. c. capacitors 115
– – conductivity 31
degree of dissociation 30
desert pollution 56
deterioration period 35
dielectric constant 67
– foil 113
– losses 37, 45, 67
difference method 4
disc winding 135
dissipation factor 67, 174
– of diphenyls 95
– – oil-filled cables 85
– – transformer oil 49, 80

distribution density 11
– function 11
dry band 57
– transformer 93, 137
duroplasts 86

earth capacitances 105
– fault winding 143, 150
edge field 99, 111
E-glass 74
elastomers 86
electric strength 1
– of gas-insulated configurations 168
electrode embedding 52
– material 53
electrolytic tank 6
electron avalanche 17, 32
electronegative gases 15, 22
electrostatic field 2
– surface force 116
energy density (in the dielectric) 111
epoxy resin current transformer 156
– – (EP) mouldings 30, 34, 37, 46, 66, 87, 91
– – support insulator 104
– – voltage transformer 147
equal area criterion 27
equipment insulator 101
erosion 45, 53, 65
error triangle 145
external control 105
– ionization 14, 16, 25
extruder 87, 90
extrusion 87, 88

ferro-resonance 152
fibre bridge, breakdown by 51
field calculation 3
– control 101, 120
– emission 25, 31, 52
– strength
– –, breakdown 9, 19, 50, 63
– –, electric 1
– –, inception 43
– –, maximum 9
filler materials 72, 90, 93
firing, sintering 76
flashover 54
flat winding 113
flow on method 60
formative time lag 26
form factor 57
Fowler-Nordheim equation 52

gap factor 28
– length 6

gas discharge 15
gases 14, 72
–, properties 168, 174
gaskets 106
geometrical characteristic 7, 159
glass 74
–, properties 174
– fibre reinforced plastics 75
gliding configuration 99
– (surface) discharge 99, 121
grading capacitors 105

hardboard 82
– tubes, attachment of 107
humidity correction 171

ignition condition 19
– time lag 27
impregnating medium 47, 84, 95, 114
impulse capacitors 116
– voltage 21, 25, 35, 98, 124, 138
– – performance of transformers 138
– voltage-time curve 26, 48
impurities
–, electrolytic 30
– of insulating liquids 47, 51
inception field strength 43
– voltage 8, 22, 43, 115, 121
– –, calculation of 43, 123
inclined boundary surface 99
incomplete breakdown 22
induction period 35, 46
inductive voltage transformer 143
industrial pollution 56
initial distribution 138
injection casting 87, 90
instrument transformer 142
– transformer error 144, 150, 153
insulated housing construction 104, 147, 158
insulating configurations 100
– materials 62
– –, inorganic 71, 174
– –, organic 79, 174
– –, properties 62, 173
– screens 52, 98
insulation 1, 62, 97
–, coordination of 13
– design 28
– material thickness 34, 112, 121
– resistance 64
insulators 100
interleaved winding 140
intermittent steps 30
inter-turn voltage 131
intrinsic breakdown 31

Index

inverted winding 135
ionic conduction 30
ionization 15, 50
− coefficient 17
− losses 45
− voltage 16
iron core 131
− cross-section (effective) 131, 154
− loss angle 132, 178
− path 131, 154

knife-edge configurations 163

law of growth of electric strength 12
layer conductivity 55
− of moisture 55, 96
− windings 135, 140, 141
lead-outs 100, 120, 122
leader mechanism 21
leakage canal 137
− current 56
lightning arrester 14
liquid insulating materials 47, 79, 81, 94, 108
− − −, properties 174
− nitrogen (LN$_2$) 48
loading error 145, 150
long-chain molecules 87
long rod insulators 76, 101
longitudinal boundary surfaces 99
loss factor 37
low-temperature breakdown 32

magnetic circuits 130
− materials, properties 177
magnetization characteristic 132, 177
magnetizing current 130, 144, 153
mean free path 15, 50
metal tank design 148, 157
mica 72
micafolium 72
micanite 72
micro-cavities 89
mineral oil 79
mobility (of charge carriers) 15, 30
multi-layer dielectrics 33, 98
− -winding transformer 135

needle electrodes 35
needle test 37
neutralization number 80
nickel-iron alloy 132, 155, 177
nitrogen 48, 71
− cushion 109
non-self-sustaining discharge 15
normal distribution (Gauß) 11

oil 79, 96, 108
− -impregnated paper 81, 82
− insulation 84, 110, 114
− -insulated devices 108
− -refining 79
− transformers 138
open-circuit error 145, 150
o-rings 106
oscillatory factor 117
overcurrent factor 155
− performance 155
overhead construction 157
− current transformer 157

paper 81
− dielectric 114
−, drying of 83
parallel-plate capacitor 111
partial capacitances 105
− discharge (PD) 9, 30
− − breakdown 35, 41, 45
− − canals 35
− −, external 23
− −, internal 42
− − configurations 42
− electrodes 103
partly crystalline materials 31, 34, 87, 90
Paschen curve 21, 168
PE-cable 89
phase angle error 145, 154
phasor diagram 145, 153
plastics (synthetics) 86
point configurations 163
polarity effect during air breakdown 24, 98
polarization 37, 67
pollution flashover 54
polyethylene (PE) 30, 87
− foil 33, 47
polymers 86
polytetrafluorethylene (PTFE) 90
polyurethane resin (PUR) 93
polyvinylchloride (PVC) 89
porcelain 76
− housing 101
potential control 105
power capacitor 118
− density (in the dielectric) 111
pre-deposited pollution method 60
pre-discharge 20, 26
pressboard 81
pressure gelation method 93
pulse corona 24

quartz 32, 68, 72
– porcelain 76
– powder filler 34, 72

radio interference 24
recombination 14
resistive layers 124
resonant inductance 149
rigid connections 106
rod gap 25, 98, 171
rod-plate configuration 24, 28, 98

salt fog method 60
– – pollution 56
saponification number 80
screening electrodes
–, external 101
–, internal 104
seals 106
secondary electrons 17, 19
self-control 105
self-sustaining discharge 15, 18
series connection of capacitor windings 113
SF_6-insulated devices 147, 158
shed profiles 101
shielding rings 138
short-circuit reactance 134
shrinkage 77
silica gel 108
silicone oil 96
– rubber 94
silicon-iron 131, 155, 177
single conductor current transformer 158
– material configuration 97
solid insulating materials 30
– properties 173
space charge 16, 22, 31
– – fields 22
– factor 131
specific conductivity 30
– heat 68
sphere gap 22, 24
spherical configurations 159
splayed flange 141
static breakdown 22
statistical time lag 25
steatite 76
streamer mechanism 20, 29
structural boundary surfaces 34
sulphur hexafluoride (SF_6) 72
switching impulse voltage 28

temperature rise calculation 109
test methods
– –, kieselguhr 60

– –, needle test 37
– –, salt fog 60
testing
– of contamination layers 50, 60
– – insulating materials 63
–, thermal stability 41, 69
thermal activation 31
– breakdown 30, 37, 50
–, global 39
–, local 41, 51
– conduction 68, 110
– conductivity 39, 68
– expansion 69
– ionization 29
–, properties 173
– transition number 69
thermoplasts 86
titanium dioxide 112
top-electrodes 102
Townsend mechanism 18
tracking strength 65
transformer board 137
– circuitry 143, 148, 152
– designs 147, 151, 156
– equivalent circuit 133
– for protection purposes 143, 152, 155
– oils 47, 108
– windings 130
– –, assembly and connection 135
– –, types 140
transmitted power 85
transverse boundary surface 98
traps 31
treeing 35, 42, 89
Trichel pulses 23
tunnel effect 31, 33
turn insulation 137

utilization factor 6, 98, 159

vacuum breakdown 52
voltage error 145
– -time area 27
– transformer 143

water trees 89
Weibull exponential distribution 11
wettability 70
withstand voltage 1, 11
winding capacitance 114
– insulation 137
wound ring core 177

X-wax formation 46, 81, 114

yoke 135

VIEWEG

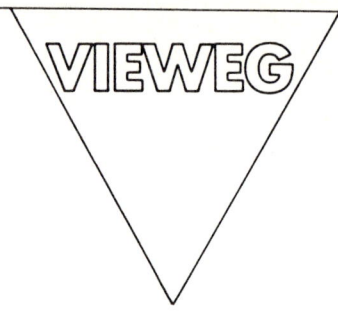

Dieter Kind
An Introduction to High-Voltage Experimental Technique
Textbook for Electrical Engineers
1978. XII, 212 pages, 181 fig. Paperback

High-voltage engineering is a field of electrical engineering deeply rooted in physics and the application of which is closely connected to industrial practice. High-voltage engineering is concerned with physical phenomena and technical problems pertaining to high voltages.

Knowledge of the properties of gases and plasmas, as well as of fluid and solid dielectrics, is of fundamental importance to high-voltage technology. Despite all progress in theoretical work, the physical phenomena in these materials are still not sufficiently understood. Hence experiment is the basis of scientific work in this field. It follows that experience in experimental work is essential for the successful treatment of numerous problems.

Recognition of this fact is the conceptual basis of this book. It is not only aimed at students of electrical engineering but also at practising engineers and physicists. Its purpose is to equip the reader for the experimental investigation of important questions in high-voltage technology. Significant practical problems of testing bays and research laboratories are described, and possible solutions indicated.

The theoretical considerations are introduced hand in hand with a detailed description of twelve experiments of a high-voltage practical. Design data of high-voltage component elements are supplemented by scaled sectional drawings to assist workshop construction.

The book thus covers the indispensable basic theory, the setup and performance of high-voltage experiments. It intends to bridge an obvious gap in the scientific and technical literature.